AF361898

The Future of Built Heritage Conservation

The Future of Built Heritage Conservation

Guest Editor

Johnathan Djabarouti

Basel • Beijing • Wuhan • Barcelona • Belgrade • Novi Sad • Cluj • Manchester

Guest Editor
Johnathan Djabarouti
Manchester School of Architecture
Manchester Metropolitan University
Manchester
United Kingdom

Editorial Office
MDPI AG
Grosspeteranlage 5
4052 Basel, Switzerland

This is a reprint of the Special Issue, published open access by the journal *Architecture* (ISSN 2673-8945), freely accessible at: www.mdpi.com/journal/architecture/special_issues/E18T3034JO.

For citation purposes, cite each article independently as indicated on the article page online and using the guide below:

Lastname, A.A.; Lastname, B.B. Article Title. *Journal Name* **Year**, *Volume Number*, Page Range.

ISBN 978-3-7258-2842-5 (Hbk)
ISBN 978-3-7258-2841-8 (PDF)
https://doi.org/10.3390/books978-3-7258-2841-8

© 2024 by the authors. Articles in this book are Open Access and distributed under the Creative Commons Attribution (CC BY) license. The book as a whole is distributed by MDPI under the terms and conditions of the Creative Commons Attribution-NonCommercial-NoDerivs (CC BY-NC-ND) license (https://creativecommons.org/licenses/by-nc-nd/4.0/).

Contents

About the Editor

Johnathan Djabarouti

Dr Johnathan Djabarouti is a registered architect and a senior lecturer of architecture at the Manchester School of Architecture, UK. His AHRC-funded research primarily focuses on the intersections between the conservation of built heritage and critical heritage theory. Since 2023, he has been a UKRI Research Associate for Historic England, where he assists the organization on the theme of intangible heritage and its relationship to place-shaping. He is the author of *"Critical Built Heritage Practice and Conservation: Evolving Perspectives"* (Routledge, 2024).

Editorial

Editorial for the Special Issue "The Future of Built Heritage Conservation"

Johnathan Djabarouti 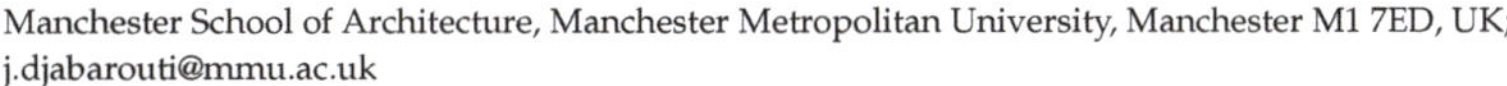

Manchester School of Architecture, Manchester Metropolitan University, Manchester M1 7ED, UK;
j.djabarouti@mmu.ac.uk

Citation: Djabarouti, J. Editorial for the Special Issue "The Future of Built Heritage Conservation". *Architecture* **2024**, *4*, 1098–1100. https://doi.org/10.3390/architecture4040057

Received: 19 November 2024
Accepted: 20 November 2024
Published: 2 December 2024

Copyright: © 2024 by the author. Licensee MDPI, Basel, Switzerland. This article is an open access article distributed under the terms and conditions of the Creative Commons Attribution (CC BY) license (https://creativecommons.org/licenses/by/4.0/).

The evolving role of built heritage in a rapidly changing world calls for a reconsideration of how we value, adapt, and protect our cultural heritage. This Special Issue of *Architecture* brings together a collection of diverse contributions that explore the complex interplay between heritage, conservation, and adaptation, considering how they inform our contemporary understanding of built heritage. The traditional understanding of heritage as tangible (physical) sites that represent authorised discourses and histories is being challenged by a postmodern conceptualisation of heritage as a dynamic and pluralistic process across space and time [1]. Together, the articles in this Special Issue provide critical insights into the future of built heritage, questioning the established methods and presenting new opportunities for more inclusive and sustainable practices.

Beginning with Laurajane Smith's established premise that "all heritage is intangible" [2], this Special Issue engages with expanding perspectives on critical heritage studies, intangible cultural heritage, contemporary conservation theory, and adaptive reuse. Evolving thinking across these themes highlights heritage as an immaterial, people-focused activity that can be constantly recreated, translated, and reborn [3]. This collection addresses the following questions: What is the relationship between the approaches to built heritage and the issues made prominent in critical heritage studies? How can architecture, conservation, and adaptive reuse strategies evolve to maintain relevance to the topical debates surrounding heritage in contemporary life? Additionally, contributions from neighbouring fields of inquiry—such as heritage management, archaeology, and cultural heritage studies—are also considered to reflect on the future directions within built heritage conservation.

Sally Stone's essay, "Notes towards a Definition of Adaptive Reuse" [Contribution 1], sets the tone by examining the evolution of adaptive reuse as a creative practice. Stone highlights how adaptive reuse transcends traditional architectural approaches, reimagining historic structures in ways that reflect contemporary cultural imperatives, such as sustainability and the focus on human experience over new construction. This emphasis on reuse and transformation forms the foundation for many of the subsequent discussions in this collection. Elizabeth Robson's article, "Assessing the Social Values of Built Heritage" [Contribution 2], explores how participatory methods can reveal the diverse and sometimes conflicting values that communities attribute to heritage sites. In focusing on Cables Wynd House in Edinburgh, Scotland, Robson demonstrates how built heritage can principally involve understanding the social significance that places hold for different community groups. Such approaches are essential for creating more people-centred, inclusive heritage conservation practices. In "Towards a Holistic Narration of Place", [Contribution 3] Youcao Ren and I delve into the complexities of conserving both natural and built elements at the Humble Administrator's Garden, China. Our research documents the tensions between the pressures of commercialisation and the need to maintain socio-cultural values, calling for more "holistic narration" that recognises both the tangible and intangible aspects of heritage. This theme of intangible cultural heritage is echoed in several contributions, such as Nigel Walter's exploration of the interplay between tangible and non-tangible heritage in "Lost in Translation" [Contribution 4], which argues for an integrated understanding of heritage that moves beyond the traditional boundaries of conservation.

The urban context also plays a crucial role in this discussion. Camilla Mileto and Fernando Vegas López-Manzanares examine "The Protection of the Historic City" [Contribution 5], outlining the challenges in balancing the needs of contemporary urban development with the conservation of historic settings. This article demonstrates how a broader view of conservation that extends beyond individual monuments is better suited to considering their urban and social contexts. Mustapha El Moussaoui's study, "Spatial Transformation: The Importance of a Bottom-Up Approach in Creating Authentic Public Spaces" [Contribution 6], underscores the value of community-driven approaches in the transformation of urban spaces. By highlighting the role of phenomenology in urban design, El Moussaoui presents a compelling case for including community narratives in the creation of public spaces, emphasising that places are constructed with authenticity when designers work with, rather than for, the community.

In a different but related vein, Lui Tam explores the transformation of Guangrenwang Temple in China, highlighting the conflict between the authenticity of heritage and modern interventions in "A Controversial Make-Over of a 'Make-Believe' Heritage" [Contribution 7]. Tam's critical examination of the "Eastern/Western" dichotomies in heritage practice adds a nuanced perspective on the global challenges in heritage while also reflecting on the fluidity of the values in heritage amidst global influences. Mazin Al-Saffar's article, "Sustainable Urban Heritage: Assessing Baghdad's Historic Centre of Old Rusafa" [Contribution 8], tackles the intricate relationship between sustainability and heritage conservation in urban contexts marked by socio-political challenges. Al-Saffar presents a mixed-methodological approach to understanding the urban heritage of Baghdad, which emphasises the importance of adopting sustainable strategies for urban evolution that resonate with both historical preservation and future city-making. In their contribution, Teresa Cunha Ferreira and colleagues examine "Joint Management Plans in World Heritage Serial Nominations" [Contribution 9], with a focus on Álvaro Siza's architectural legacy. The discussion of joint management plans for World Heritage sites emphasises the significance of collaboration and shared responsibility in managing complex heritage sites, offering insights into the future of managing serial nominations as a rising trend in global heritage governance. Lastly, Ataa Alsalloum's "Building Home in Exile: The Role of Intangible Cultural Heritage, Crafts, and Material Culture Among Resettled Syrians in Liverpool" [Contribution 10] shifts our focus to heritage at the personal scale. Alsalloum examines how displaced Syrians in Liverpool, the UK, use intangible cultural heritage as a tool to create a sense of home, blending traditional crafts and practices with their new environments. This intimate portrayal of heritage speaks to the resilience of cultural identity and the role of intangible heritage in shaping a sense of belonging within the diaspora.

These contributions articulate an understanding of heritage that is increasingly multi-layered, dynamic, and inseparable from contemporary social, environmental, and cultural concerns. The future of how we engage with built heritage will no doubt move beyond a focus on material fabric and decay prevention towards a more fluid, adaptive, inclusive, and socially engaged approach that foregrounds the diverse values people imbue their physical environments with. This aligns with critical heritage discourse that regards heritage as a pluralistic, ongoing process—a social performance that reflects and (re)shapes community identities. The editor hopes that this Special Issue will inspire readers to actively engage in future-focused heritage processes, ensuring heritage continues to evolve, resonate, and thrive in our rapidly changing world.

Funding: This research received no external funding.

Conflicts of Interest: The author declares no conflicts of interest.

List of Contributions

1. Stone, S. Notes towards a Definition of Adaptive Reuse. *Architecture* **2023**, *3*, 477–489. https://doi.org/10.3390/architecture3030026.
2. Robson, E. Assessing the Social Values of Built Heritage: Participatory Methods as Ways of Knowing. *Architecture* **2023**, *3*, 428–445. https://doi.org/10.3390/architecture3030023.
3. Ren, Y.; Djabarouti, J. Towards a Holistic Narration of Place: Conserving Natural and Built Heritage at the Humble Administrator's Garden, China. *Architecture* **2023**, *3*, 446–460. https://doi.org/10.3390/architecture3030024.
4. Walter, N. Lost in Translation: Tangible and Non-Tangible in Conservation. *Architecture* **2023**, *3*, 578–592. https://doi.org/10.3390/architecture3030031.
5. Mileto, C.; López-Manzanares, F.V. The Protection of the Historic City: The Case of the Surroundings of the Lonja de La Seda in Valencia (Spain), UNESCO World Heritage. *Architeture* **2023**, *3*, 596–626. https://doi.org/10.3390/architecture3040033.
6. El Moussaoui, M. Spatial Transformation—The Importance of a Bottom-Up Approach in Creating Authentic Public Spaces. *Architecture* **2023**, *4*, 14–23. https://doi.org/10.3390/architecture4010002.
7. Tam, L. A Controversial Make-Over of a "Make-Believe" Heritage—The Transformation of Guangrenwang Temple. *Architecture* **2024**, *4*, 416–444. https://doi.org/10.3390/architecture4020023.
8. Al-Saffar, M. Sustainable Urban Heritage: Assessing Baghdad's Historic Centre of Old Rusafa. *Architecture* **2024**, *4*, 571–593. https://doi.org/10.3390/architecture4030030.
9. Ferreira, T.C.; Freitas, P.M.; Cruz, T.T.; Mendonça, H. Joint Management Plans in World Heritage Serial Nominations: The Case of Álvaro Siza's Modern Contextualism Legacy. *Architecture* **2024**, *4*, 820–834. https://doi.org/10.3390/architecture4040043.
10. Alsalloum, A. Building Home in Exile: The Role of Intangible Cultural Heritage, Crafts, and Material Culture Among Resettled Syrians in Liverpool, UK. *Architecture* **2024**, *4*, 1020–1046. https://doi.org/10.3390/architecture4040054.

References

1. Djabarouti, J. *Critical Built Heritage Practice and Conservation—Evolving Perspectives*; Routledge: Oxon, UK, 2024. [CrossRef]
2. Smith, L. *All Heritage Is Intangible: Critical Studies and Museums*; Reinwardt Academy: Amsterdam, The Netherlands, 2011.
3. Fouseki, K. *Heritage Dynamics: Understanding and Adapting to Change in Diverse Heritage Contexts*; UCL Press: London, UK, 2022.

Disclaimer/Publisher's Note: The statements, opinions and data contained in all publications are solely those of the individual author(s) and contributor(s) and not of MDPI and/or the editor(s). MDPI and/or the editor(s) disclaim responsibility for any injury to people or property resulting from any ideas, methods, instructions or products referred to in the content.

Essay

Notes towards a Definition of Adaptive Reuse

Sally Stone 

Manchester School of Architecture, Manchester M1 7ED, UK; s.stone@mmu.ac.uk

Abstract: This essay will discuss the evolution of writings about adaptive reuse. The architectural practice is as old as the buildings themselves, yet it has scarcely been discussed or even recognised until relatively recently. The essay will document the varied influences that informed the early publications (the first from 1976). The lack of easily available material (that is, books and documented buildings) meant that pioneering writers had to draw upon other sources—those beyond established architectural discussions. Therefore, these early authors were not limited by the strictures of an already established subject but were able to collate information from a variety of sources. Thus, adaptive reuse draws upon a collage of different sources, many beyond pure architecture, including installation art, fine art, curation, interior design, and urban design. Inevitably, as the subject moves from the periphery of architectural practice towards the middle ground, the number of publications has increased. This diversity has provided the subject with a greater scope, supporting the acknowledgement of the importance of technology, sustainability, and conservation in addition to ideas of heritage and culture, while also allowing for a much less Western-centric focus.

Keywords: adaptive reuse; interior architecture; installation art; remodelling; literature; publications

Citation: Stone, S. Notes towards a Definition of Adaptive Reuse. *Architecture* **2023**, *3*, 477–489. https://doi.org/10.3390/architecture3030026

Academic Editor: Avi Friedman

Received: 7 July 2023
Revised: 19 August 2023
Accepted: 23 August 2023
Published: 28 August 2023

Copyright: © 2023 by the author. Licensee MDPI, Basel, Switzerland. This article is an open access article distributed under the terms and conditions of the Creative Commons Attribution (CC BY) license (https://creativecommons.org/licenses/by/4.0/).

1. Introduction

Fragments of a vessel which are to be glued together must match one another in the smallest details, although they need not be like one another. In the same way a translation, instead of resembling the meaning of the original, must lovingly and in detail incorporate the original's mode of signification, thus making both the original and the translation recognizable as fragments of a greater language, just as fragments are part of a vessel.

Walter Benjamin [1] (p. 260)

For such a long-established and deeply entrenched subject, adaptive reuse has a remarkably short history. It is a practice that stretches back to almost the first constructed buildings themselves, for structures have perpetually been altered to accommodate the needs of their different occupants [2] (p. 114), and yet it has continually lacked the written theoretical and historical recognition of new-build architecture. This could be based upon the prejudice inherent in the concept that interior design (as the practice was once called) was regarded as a respectable profession for women, combined with the lack of perceived worth in adaptive reuse within the modernist and late-modernist world. However, the agenda of the 21st century has ensured that adaptive reuse is beginning to be accepted as a professionally relevant and creative method of developing the built environment. This builds towards this century's environmental urge to adapt and transform combined with the need to build human experiences, rather than construct new things. The current mantra "reuse, reduce, recycle" is an indication of this massive shift in attitude. Adaptive reuse is now seen as one of the most significant issues within the architectural profession.

The act of working with the already-built implies compromise, it suggests that the designer, rather than imposing their own vision upon a specific place, must first understand the agenda of the building before presuming to change it. This implies negotiation, agreement, and conciliation. But adaptive reuse is also transgressive; it undermines the primacy of the original architect, makes secondary his or her agenda, and overlays this with new meaning.

The recognition of the practice of adaptive reuse as an appropriate architectural approach is still so new that an exact title has yet to fully emerge (also, an exact definition). Adaptive reuse does seem to be evolving into the settled term for the subject, however, the practice can also be referred to, among other options, as: interior architecture, remodelling, building reuse, retrofitting, conversion, adaptation, rehabilitation, reworking, refurbishment, or, sometimes, especially in North America, repurposing. These are reverential terms, possibly transgressive or subversive, but they do not exhibit overt authority, so it is interesting that the website 'Building on the Built' refers to the practice as 'interventional work'; these are both assertive words that seem to elevate the approach into a much more proactive, definite, and less deferential activity.

Certainly, there are many terms to describe the process; Graeme Brooker and Sally Stone's ReReadings described it as *interior architecture* [3], while Fred Scott and Philippe Robert both use the single word *adaptation* [4,5]. Even as late as 2022, there was still ambiguity about the title and the definition; Francesca Lanz and John Pendlebury in a 2022 essay, called Adaptive Reuse: A Critical Review, declared that... *"there is no common and shared agreement on what adaptive reuse precisely is and what it entails"* [6]. Fred Scott describes the subject as *alteration*, which he defines as the "mediation between preservation or demolition" [4]. Bie Plevoets and Koenraad Van Cleempoel propose that it is *altering existing buildings for new or continuous use* [7]. Johannes Cramer and Stefan Breitling suggest that it is *architecture within existing built contexts* [8]. James Douglas: *Any work to a building over and above maintenance to change its capacity, function or performance* [9]. The ICOMOS definition is: *Adaptation means the processes of modifying a place for a compatible use while retaining its cultural heritage value. Alteration processes include alteration and addition* [10]. While the Burra Charter says that *Adaptation means changing a place to suit the existing use or a proposed use* [11]. Sally Stone's monograph UnDoing Buildings suggests that adaptive reuse is described as utilizing *strategies that are applied not as a reaction but in anticipation* [12]. Phillipe Robert elaborates to suggest that it is the *story of the successive layers, of the reshaping of monuments, and of the additions that bear testimony to each succeeding age* [5], and Frank Peter Jager simply states *work with existing buildings* [13].

> *The past provides the already written, the marked 'canvas' on which each successive remodelling will find its own place. Thus the past becomes a 'package of sense' of built up meaning to be accepted (maintained), transformed or suppressed (refused).*
>
> Rodolfo Machado [14]

2. A Review of Sources

This is an opportunity to dwell upon the evolution of the subject and to discuss the literature that has informed the approach, plus the buildings, architects, artists, and installations that have contributed towards that formation. Given that adaptive reuse has evolved into one of the predominant aspects of architecture in the 21st century, it would be interesting to approximately divide this study by the turn of the millennium.

3. Common Link

The common link throughout this discussion is the connection with place. The idea that the authenticity of place, the reality of a tangible situation, and the sensory connection with the actual physical certainty of somewhere substantial and quantifiable can not only create a connection with the past, but can also generate a new future. In an anxious world of continual surveillance, with virtual realities that are not necessarily real and truths that are not completely true, this connection with the actual physical situation of a definite place provides a sense of certainty that is often not readily available elsewhere. Adaptive reuse creates a real connection with place. The relationship is real and tangible, it is authentic and contains certainty.

Adaptive reuse responds to the situation to which it is directly connected. This is both a tangible physical connection with the material reality of the environment, but also with the intangible collection of forces that formed it. Whether these are cultural, climatic or geographical, man-made or natural, they are elements that comprise the situation of the place, they inform its character and the way in which people react to it. The crisis within the contemporary city means that continued horizontal development can no longer be supported, but the built environment needs to build in on itself, to be more dense, more productive, more resilient. Buildings, situations, and neighbourhoods are in a continual state of flux, they are altered, updated, maybe rehabilitated, but rarely do they exist in a state of scarification. This incessant renewal is an opportunity to accommodate the needs and aspirations of those who occupy the place, hopefully before they even realise that they need the change. The connection with place is exemplified by how intricately these elements are linked together, how this connection can create a ripple through the continuity of existence, and how it can build a better future.

4. Before the Millennium: A Collection of Texts, Buildings, Interiors, and Installations

There are two factors that tie together this collection of texts, buildings, interiors, and installations. The first is that intrinsic relationship between the built form and the environment that it inhabits. Contextualism (which is a design tool/approach rather than a style) connects all the sources discussed. Adaptive reuse projects enjoy the double dialogue of the context of surrounding area of occupation, plus the conditions of the host building.

The second factor is the attention to detail. The original building is an intense collection of tangible and intangible elements, an assembly of real and virtual parts that gather together to form a coherent image of the structure, and it is to these that the interventions of reuse render respect and respond. To be able to alter a building, the designer needs to develop an intimate understanding of individual parts. The list of areas of understanding is long: materials, methods of construction, structural system, rhythm of spaces, position of the openings, circulation, and many more, all of which contribute to the physical conditions of the existing building. This is combined with more intangible components connected with the culture of those who first constructed the building, those who occupied it, and those who will occupy it. To develop an intense dialogue between old and new, the designer needs a personal relationship with the old and the new on an intimate scale, for these are not abstract buildings on greenfield sites, but they are an exquisite uniting of two allied but not identical individual surfaces that flex and deform, soften and align, to create a union of convergence.

5. Significant Publications about Adaptive Reuse

By the beginning of the 21st century, there were books full of case studies, books that discussed the practicalities of the subject, picture books that tickled the surface of the subject, books that just about included the area as a periphery to the focus of the discussion, and essays that touched on it, but just two publications that systematically analysed and discussed the process, approach, or methodology of adaptive reuse: Machado's *Old Buildings as Palimpsest* and Robert's *Adaptations*.

Probably the most relevant is Rodolfo Machado's *Old Buildings as Palimpsest* [14]. The U.S.A. journal *Progressive Architecture* published the four-page essay in 1976; initially it was relatively unknown but over the last half-century it has become a recognised approach and indispensable source. Machado's use of the palimpsest as an analogy for the process of adaptation perfectly describes the pluralism inherent within the approach. The essay also introduces the concept of "form following form". Machado declared that "... *the form/form relationship is the primary consideration within remodelling activity*" [14]. This turned the prevalent mantra "form follows function" on its head. The idea that the influence of the enclosing buildings is so great that it becomes the primary driver for the methodology of reuse was revolutionary and far from the prevailing idea that the relationship between old and new was secondary to the proposed function and the ego of the architect.

The other important publication was *Adaptations: New Uses for Old Buildings* by Philippe Robert [5]. It was initially published in the *French Architecture Thematic Series* by Editions du Moniteur in 1989 and was translated into English by Murray White and published by Princeton University Press in 1991. The book is radical, it broke new ground, it was progressive, and, as the front runner, very important. It is organised in a simple tri-part order: 1. Introduction; 2. Detailed case studies illustrated with photographs, measured drawings, and sketches; 3. A small catalogue of recently constructed landmarks.

The very short introduction is rich and powerful. Conversion, the author declares, can be considered as a *"normal architectural practice"* [5] (p. 6)—a distinctly early proclamation for what is now a ubiquitous approach. Later in the same paragraph, Robert asserts that the renewed awareness of the history of architecture includes the *"history of buildings that have been altered"*. This, of course, coincides with the post-modern ideas of the return of history, the importance of the individual, the embracement of pluralism, and the search for eclecticism.

Within the density of the introduction, Robert also discusses the idea of the palimpsest as a metaphor for adaptive reuse. Machado is not listed in the bibliography, but this idea, and that of the form–form relationships, are analysed. In fact, Claude Soucy is listed as the source for this and quoted thus: *"Out of the encounter between old envelope and new requirements and means, a unique object will be born-one which is no mere juxtaposition, but a synthesis from the point of view of both construction and architecture"* [5] (p. 9).

An innovation in the intense opening chapter is the classification of the different approaches to adaptation. Robert lists seven strategic types of approach: building within, building over, building around, building alongside, recycling materials, adapting to a new function, and building in the style of. This inventory is intriguing and, yet, also unwieldy and messy—the taxonomy seems too ambiguous; it lacks focus and appears incomplete. Despite the creative originality in the classifications, it is difficult to place a number of key buildings within any category. For example, a more exact group is needed to house Scarpa's masterpiece Castelvecchio Museum (Verona, 1956–1973) and his ethereal Querini Stampalia Foundation (Venice, 1963), a group that would recognise the scraping away of parts of the building and the addition of a series of new elements. The Tate Modern (Herzog and de Meuron, London, 1999) is difficult to classify, as is the Irish Film Centre (O'Donnell and Tuomey, Dublin, 1992), which is the conversion of nine connected buildings.

There are a couple of other books that are focussed upon adaptive reuse, and, certainly, at a time of scarcity, were significant, but this has faded over time. The 1989 publication *Re/Architecture* by Sherban Cantacuzino [15] is a beautifully illustrated book with over 50 case studies. It contains six chapters and is organised by the function of the original building, so, for example, the first chapter examines public buildings, and the first case study is the conversion of the Helsinki City Hall for multi-use purposes, and the second is the transformation of the Gare d'Orsay in Paris into a national museum. Each chapter is prefaced by a well-informed and accessible introduction. The main introduction to the book discusses the importance of the stock of existing buildings as an opportunity for urban regeneration, but also useful for *"sound economic, social and ecological reasons"* [15] (p. 9). A far-sighted prophesy indeed! Kenneth Powell's *Architecture Reborn: The Conversion and Reconstruction of Old Buildings* [16] takes a similar approach to the structure of the book but uses the transformed function rather than the original use as the subject for each chapter. The book, as would be expected from someone with such a reputation, is very well researched and engagingly written; it is big and dense, with well-produced photographs and supported by architectural drawings. Powell's introduction, as Cantacuzino, begins with an historical survey, but ends with a call to arms. *"The issue is no longer about new verses old"*, he declared, *". . . but about the nature of the vital relationship between the two."* The introduction concludes with an assertive quote from David Chipperfield: *"We must inhabit an ever-evolving present, motivated by the possibilities of change, restricted by the baggage of memory and experience"* [16] (p. 19).

6. Other Publications That Discussed a Contextual Methodology

Post-modern pluralism, new urbanism, and contextualism all played an important role in the rise of a specific adaptive reuse theory. There were a number of highly influential books; these did not discuss adaptive reuse per se, but certainly included it among the searching ideas for a new urbanism. Seminal publications, such as Jane Jacobs' *The Death and Life of Great American Cities* [17], Colin Rowe and Fred Koetter's *Collage City* [18], Aldo Rossi's *Architecture of the City* [19], Robert Venturi's *Complexity and Contradiction in Architecture* [20], Michael Graves' guest editorship of the *Roma Interrotta* project [21] edition of *Architectural Design*, and Thomas Schumacher's short essay *Urban Ideals and Deformations* [22] all promoted the idea of the city as an eclectic mix of old and new that could together create a progressive and harmonious future. All of these ideas dealt with the development and redevelopment of the existing built environment, and, so, were easily extended to the adaptation of existing buildings.

Graves's editorship of *Roma Interrotta* of 1979 documented the project that was invented and developed by Piero Sartogo, which took the breath-taking Nolli Plan as inspiration. Sartogo asked 12 prominent architects to reimagine it, each taking a proportional section of the great drawing as both the starting and the finishing points. So, the drawing was complete at the beginning of the process, and whole, once again, at the end, but, as a palimpsest, it had been rubbed away and redrawn during the course of the project. The Roman Interventions interrupted the drawing; they did not obliterate the grain of the city, the organisation of the streets and squares, the position of the buildings, and the arrangement of the interiors, instead, the architects worked with these attributes, producing what was then a radical fusion of old and new.

Colin Rowe was one of the guest contributors to the reimagined Nolli Plan who, together with Fred Koetter, had published *Collage City* just a year earlier. This narrative discussed the crisis within the modern city, the problems connected with the obliteration of history and the need for a more contextual approach to architecture. The book, which starts as a methodical undoing of the prevailing attitudes towards architecture and urban design, continues as a call to arms for a new approach, and ends as a handbook of inspirational approaches to guide the way forward.

Rowe and Koetter discuss such romantic suggestions as the apotheosis of the collision, the search for bricolage, and the reconquest of time. A significant discussion is the comparison between le Corbusier's monumental *Unité d'Habitation* and Giorgio Vasari's *Uffizi Palace in Florence* [18] (p. 69). One is the inverse of the other, so, while the Unité is a solid monolith, so the Uffizi is a void. Thus, the area of land surrounding the Unité is deformed to accommodate the regular building, and, conversely, the building surrounding the void, or Vasari's Corridor (as it is known), is deformed to accommodate the space, so undermining the modernist ideas of the primacy of form and opening up the possibilities of building in and around existing structures.

The idea of *Contextualism: Urban Ideals and Deformations* was further explored by one of Rowe's students, Thomas Schumacher [22]. He treads very much the same path as his tutor, but in nine intense pages that call for some sort of middle ground between an artificial incarnation of the past and the brutalising and dominating system of modernism [22] (p. 297). An ideal form can exist as a fragment *"collaged"* into an empirical environment [22] (p. 301). Contextualism, he asserted, is a design tool that could be abstracted to any given situation. Kate Nesbitt who collected the essay her edited collection: Thoerizing a New Agenda for Architecture [23] (p. 294), recounts that Schumacher's recollection was that Contextualism is a conflation of Context and Texture. The term he suggested, was first used by Steven Hurtt and Stuart Cohen.

Venturi and Scott Brown's *Complexity and Contradiction in Architecture* [20] uses historical precedents to propose a methodology for moving forward; an attitude that suggests that everything is valid, that there is a need to move away from the tabula rasa approach, and, even more so, away from the primacy of the monumental volume. The opening section, entitled Nonstraightforward Architecture: A Gentle Manifesto, called for elements that are "*...hybrid rather than pure, distorted rather than straightforward, ambiguous rather than articulated...*" [20] (p. 16). (Robert Venturi is listed with the authorship of the book, but Denise Scott Brown's contribution is now so recongnised that she is normally credited as co-author.) Especially relevant to this discussion is chapter 9, on the importance of the interior, and the understanding that the exterior and the interior could have different personalities; that the interior is much greater that the mere consequence of the containing exterior walls. They railed against the modernist orthodoxy of the continuity between the inside and the outside, that one should slip easily into the other to create a "*oneness*" [20] (p. 301). (This attitude later caused the great interior theorist Fred Scott to suggest that "*the interior had escaped from the building*" [12] (p. 15)).

Venturi and Scott Brown supported the idea that the exterior and the interior of a building could be different, and this separation, he argued, emphasised the identity of both. They reasoned that contradiction may be further emphasised through the use of detached linings, which can leave spaces between the structure and the interior, thus providing opportunity for interpretation. They were not explicitly discussing adaptive reuse but more providing the springboard for further consideration. This exploration of difference was an incentive for remodelling. Importantly, this is greater than a book about urbanism, it is about the comfort of enclosing space rather than the significance of epic building, a pursuit of modesty, about the understanding of how a collection of intricate details can create a greater whole, of how the environment of the already-built could provide the impetus for future development, and how all of these could appear to have always been there but are so obviously of the now.

Another significant publication that opened as an attack on the principles and aims that have shaped modern, orthodox city planning and building, then evolves into a manifesto for an exuberant and diverse city, is Jane Jacobs' *The Death and Life of Great American Cities* [17]. Her far-sighted call for the new to mingle with the old, for building to address the street, for places to serve more than just one primary function, and for density of population [17] (pp. 150–151), is now acknowledged as an astute recognition of how to address the 21st century concern with sustainable population growth in cities. Some 60 years after publication, densification conducted through the adaptive reuse of the stock of existing buildings is the established approach to development.

These texts were important to the development of a methodology for adaptive reuse, they regarded the built environment as an evolving situation of discourse, and the ideas developed and discussed were as relevant to individual buildings as they were to larger urban environments.

7. Art, Architecture, and Design

As important as the publications was the work of specific architects and designers who pursued a contextual approach in their work combined with a love of heritage and history; they also searched for narratives and fables, and wrapped this in a post-modern sensibility. Visual people habitually spend longer looking at the pictures than reading the words, so buildings displayed in such journals as *Blueprint* or *Journal of Interiors*, combined with visits to the places, were often more important than the texts that discussed them. Architects and designers included Carlo Scarpa, Group 91, O'Donnell and Tuomey, Hans Hollein, John Outram, Nigel Coates, James Stirling, Vittorio Gregotti, Aldo Rossi, Robert Venturi and Denise Scott-Brown, Ron Herron, Aldo Van Eyck, Rafael Moneo, Coop Himmelblau, Ken Belly, David Chipperfield, Memphis...

Carlo Scarpa is regarded as the master of adaptive reuse, yet his work was often acknowledged as lacking architectural intent. The practice of adaptive reuse has long been seen as having limited worth and beneath the interest of many architects. Even as late as 1993, Richard Murphy, in his highly detailed and intense discussion of the Fondazione Querini Stampalia Foundation (Venice, 1963), questioned the veracity of the design and asked whether Scarpa's work was "*merely interior design*" [23] (p. 3).

And yet, young architects were beginning to establish a reputation with such projects. David Chipperfield's early shop interiors (for example, the Issey Miyake boutique, Sloane Street, London, 1985) were formative little projects that combined careful craftmanship with an exploration of complex spatial relationships. Hans Hollein created a series of daring, yet refined, individual shops in Vienna (for example, the Retti Candle Shop, 1966, or the Schullin Jewelery Store, 1974). These long, narrow stores that appear slotted into the available spaces are exquisitely executed interiors—as would be expected in the birthplace of the Secessionist movement. The great documenter of post-modernism, Charles Jenks, wrote a rapturous review of Hollein's early work: "*So much design talent and mystery expended on such small shops would convince an outsider that he had at last stumbled on the true faith of this civilisation*" [24] (p. 32).

The radical post-modern architect John Outram is recognised for the ground-breaking Pumping House in the Isle of Dogs (London, 1986). Outram's buildings are borne from ancient myths and modern parables and invoked the inherent romance of Claude Lorraine's landscapes; the pumping station in the Isle of Dogs conceptually contained columns that penetrated hundreds of feet through the mud and silt to connect with the bedrock, while the roof of the Kensal Road housing swoops gracefully from above to gently land upon the building. He also completed the transformation of an ordinary two-storey 1960s concrete-framed office block into an articulate yellow brick-clad building which was seemingly supported by great bulbous columns with flaming capitals. But it was cleverer than a mere cosmetic revamp. All of the services were diverted into ducts hidden within the fat columns, and those that did not contain such facilities performed other useful services—the coffee machine, the filing cabinets, the fire extinguishers.

The Temple Bar Framework Plan in Dublin by Group 91 [25] (Dublin, 1996) was equally influential. The substantial area next to the River Liffey had been earmarked for a huge bus station; in fact, in 1977 Skidmore, Owings & Merrill Architects produced a scheme for a great spiraling monolith to completely fill the site. When, years later, this proposal was abandoned, the city council held an architectural competition for the complete neighbourhood. The winning project proposed to regenerate the area through the construction of a series of cultural buildings, which would tuck into the urban grain of the area, thus allowing the natural rhythm of the place to be retained. The scheme proposed a mixture of new buildings and adaptations, the most notable being the Irish Film Centre. O'Donnell and Tuomey were individually responsible for this amalgamation of nine different existing buildings set deep within the city block.

Another seminal adaptation and, perhaps, the last to mention here is the Haçienda (Manchester, 1982)—once described as the "*most famous night club in the world*". Ben Kelly's joyful, post-industrial, post-modern approach to adaptive reuse has proved to be absolutely revolutionary, and his paradigm-changing design for the interior of the nightclub has become part of a powerful cultural legacy rooted in both the city's and the era's industrial aesthetic; it has proved to be internationally influential.

It is also important to discuss the influence of installation artists to the development of adaptive reuse. Artists can experiment with existing buildings and spaces without the pressure of the needs of the end users and the exacting regulations connected with construction, therefore, they are often in the position to push ideas further and more quickly than the architect or designer is able.

The Gordon Matta-Clark retrospective at the Serpentine Gallery in London, 1993, [26] was a timely and powerful exploration of the impact that considered dissection can have upon existing buildings. Matta-Clark cut holes in buildings, whether to create connections that did not previously exist, to reveal unfound associations, and in one piece, Splitting, he actually cut a timber house in half, to expose the flimsy insubstantial nature of the structure and, maybe, also of inhabitation itself. The exhibition caught the mood of many architects and designers of the time who were beginning to question the dominance of new buildings when perfectly good strong and useful ones still existed, of the removal of built heritage, and the prevailing lack of legacy that resulted. The other important aspect of Matta-Clark's work was the conceptual idea of the subtraction of material. This was tantamount to an anti-heroic architectural move; it was exactly the opposite of the progressive and productive monument to the exaltation of the architect.

There was another installation of equally massive impact the same year as the exhibition; House by Rachel Whiteread (London 1993). This installation uncovered the actual space within the interior of a single house in a soon-to-be-demolished terrace in London. Whiteread used the structure as a mould to create a three-dimensional representation of the interior of the rooms by spraying the inside of the exterior walls with concrete then removing the walls, thus leaving the insides exposed. This included the reverse of the mouldings around the doors and windows, the reverse of the windows, and, significantly, the patina of time and use on the walls themselves.

The artists, although a generation apart, were equally radical. They questioned the substance from which buildings were constructed, and, by extension, the basis of the society that constructed those buildings. The exposure of the insides of the House was deemed to defile the people who had once lived there, while Splitting was seen as a comment about the insubstantial lives of those who occupied it. Despite their shock appeal (and, by the end of the 20th century, it was getting very difficult to shock people) their work was beautiful, poised, and knowing—about architecture, structure, balance, and life.

There are other artists who were also important, these include Cornelia Parker, whose installation Cold Dark Matter: An Exploded View is the exposed violence of a detonated shed. Again it uses an existing structure and the beauty of the resultant installation creates impact and questions the strength and permanence of the built structures around us. Robert Irwin (*There is No There There until You See There There*), James Turrell who created exquisitely clever installations with pure light, Richard Wilson whose installation of a huge treacherous tank of thick dark reflective oil in the Sachi Gallery gave the space an ambiguous shape and size, and Alison Turnbull, whose manipulated images were generated by seemingly randomly discovered architectural drawings that were then expanded, revised, changed and subjected to alterations that, like the palimpsest, retained the essence of the original, but created a completely new proposal. These artists explored existing spaces and forms, then attempted to heighten the impact of these given places through considered interventions.

8. After the Millennium: An Examination of the Canon—Books about Adaptive Reuse

These conditions generated a collection of publications that have begun to create a canon of thought about adaptive reuse. Given the relative youth of the subject, the books are spare and focussed, but it is interesting to observe that, as the 21st century progresses, how the breath is being discovered. Over the last 20 years, the number of books specifically about adaptive reuse has proliferated. The majority of these can be easily divided into two categories: those that make extensive use of case studies to illustrate themes or processes (Frank Peter Jäger, Christian Schittich, Graeme Brooker and Sally Stone, David Littlefield and Saskia Lewis, Johannes Cramer, and Stefan Breitling), and those that carefully build the argument through a series of illustrated discussions or chapters (Fred Scott, Lilian Wong, Sally Stone, Bie Plevoets, and Koenraad Van Cleempoel).

9. Case Study Books

It is inevitable that the case study books should use a similar organisational approach to those published before the turn of the millennium, but the focuses of the studies differ; from quite technical explorations, through poetic interpretation, to books that shout about the urgency of the situation. The system of classification, rather like a translation, is always partial and emphasises the interests and obsessions of the author(s). This subjective process of taxonomy is determined by the culture and experiences of the individual(s) who make the selection, thus, there are both different selections of buildings and different interpretations of the chosen buildings. Keith Jenkins explains that the basis of this emphatic system of interpretation are the morals imposed by contemporary society, and that "*. . .given that interpretations of the past are constructed in the present, the possibility of the historian being able to slough off his present* *to reach somebody else's past on their own terms looks remote"* [27] (p. 40). To extend this further, the preoccupations of the authors guide the taxonomic process. This is doubly complicated, as the process of bringing a building from a past existence into the present can be seen as a work of translation (Scott, Stone, Van Cleempoel), as the inspirational equivalent to transcribing from one language to another.

It is within the discussion of the introductions that the differences are revealed. Jäger, whose criteria for case study selection is architectural quality, describes the process as *"A Gift from the Past"* [28] (p. 11), Schittich, whose selection is deliberately optimistic, describes it as *"Creative Conversions"* [29] (p. 9), Cramer and Breitling pursue clarity—in both the intellectual process and construction techniques [8] (p. 9), Stone and Brooker pursue an architectural approach [3], while Littlefield and Lewis place adaptive reuse among a great artistic tradition of decay and rebirth [30] (p. 15).

Brooker and Stone's *ReReadings: The Principles of Interior Architecture and the Reuse of Existing Buildings Volume 1* (2004) was probably at the vanguard, but there are significant books not far behind. The book, which builds upon a synthesis of the pre-21st century texts and precedents, was at the forefront of an oncoming movement that placed much greater emphasis upon the already-built, that valued history and heritage, that used a post-modern sensibility to create a new future that learnt from the past but, equally, considered the need and aspirations inherent in the future. But, unlike much of the previous literature, *ReReadings* presented a methodology for the future of the already-built. *ReReadings* assembled the collection of impulses and arranged them in a comprehensible order that rendered the process accessible to all involved. It set this out in easy stages, and, so, a process that had previously been seen as slightly impenetrable, complicated, and difficult to read was rationalised.

When the first volume was published, it broke new ground, it proposed ideas that, although part of a continuity, were quite radical. The urgency of the book combined with the lack of easily available information means that it has a very Western focus. There were few case study projects beyond Europe and the U.S.A. The reflection made possible by the decade and a half between the publication of the two volumes in the series allowed for a more relaxed and inclusive approach to the selection of case studies in *ReReadings: The Principles of Interior Architecture and the Reuse of Existing Buildings Volume 2* (2018). Diversity is demonstrated through the selection of the projects; this was an opportunity to expand the geographical areas discussed. The focus is still predominantly European, but projects in Malaysia, China, Japan, Taiwan, Brazil, and Australia are also presented. Discussions of sustainability, digital and other methods of occupation, plus a much less Western-centric selection of case studies, pushes the argument beyond the normal bounds of architecture and interiors, and embraces many of the cross-disciplinary and diverse aspects of the subject.

Building in the Existing Fabric: Refurbishment, Extensions, New Design (2003), edited by Christian Schittich, expresses very similar sentiment, that a turning point in the attitude towards existing buildings has been reached and conversions are, the author declares, the *"New Normal"* [29] (p. 9). He suggests that conversion and renovation are no longer seen as a *"necessary evil"*, that things have changed, and the process has become one of the *"most creative and fascinating tasks in architecture"*. The introduction acutely states that *"For a long time, Carlo Scarpa's refurbishment of the medieval Castelvecchio in Verona (1956–1964) was considered the benchmark for all creative conversions"* [29] (p. 9), and also suggests that none of its vitality has been lost. There are 24 good examples of adaptive reuse described, and each is accompanied by good photographs and detailed drawings. In *Architectural Voices: Listening to Old Buildings* (2007), David Littlefield and Saskia Lewis, again, catalogue a well-researched and insightfully described collection of case studies which are prefaced by a nostalgic introduction that almost wistfully looks for traces of romance within the history and patina of the existing structure. The bibliography reinforces the lack of available literature in these first few years of discovery, with just one book about adaptive reuse [30]. Frank Peter Jäger's *Old & New—Design Manual for Revitalizing Existing Buildings* [28] utilises three groupings: addition, which discussed new elements within or around the existing; transformation, which represents a change of appearance; and conversion, which denotes a change of use. The case studies are well illustrated, the discussions certainly have technical depth, and there is an emphasis on projects that contain aspects of the socio-political, but the classifications do seem somewhat arbitrary. Johannes Cramer and Stefan Breitling in *Architecture in Existing Fabric: Planning, Design Building* (2007) [8] use case studies grouped into chapters each with in-depth discussion. The order of these is ingeniously dictated by the design process itself, so the chapters are, from the beginning, Planning Process, then Preparatory Investigations, followed by Design Strategies, Detail Planning, Building Works, and concluding with Sustainability, which, if the truth be known, and however important the subject is, does feel a little bit like an add-on at the end. The series, *Basics Interior Architecture: An Approach* [31–33], again by Brooker and Stone, also fits into this chapter-driven illustrated case study category. The methodology for these is informed by the analysis of the existing situation combined with a developed approach to the changes proposed by the architect. The process of reading an existing building can be divided into three basic categories: *Form and Structure* (2007, reprinted 2016), *Context and Environment* (2008), and *Elements and Objects* (2010). Suzie Attiwill describes the organisation of Brooker and Stone's *ReReadings* (and this could also apply to all of the books in this section) as rather *"like a curated exhibition, various examples are selected to illustrate each category"* [34].

10. Books That Build the Argument through Chapters

A number of the books divide their argument into chapters. These books embrace the dramatic change in attitude towards adaptation—from the architects and designers that intervene within the buildings, the developers who have begun to appreciate the value inherent within the already-built, to the legislators who have realised the importance of continuity to the mental and physical wealth of a community.

Within these publications, the chapters are generally stand-alone and can be read as individual discussions, however, the books do tend to construct this as a narrative or journey; so, Stone's *UnDoing Buildings* [12] starts with the strategic approach, advances through peripheral, yet influential, issues such as conservation and installation art, towards resolution within the details. Scott's *On Altering Architecture* [4] commences by constructing the case for reuse, and each chapter reinforces this before the book concludes with some resolutions. Wong's book *Adaptive Reuse: Extending the Lives of Buildings* [35] contains 15 informed chapters that each tackle a different aspect of the subject but, equally and individually, each makes the case for reuse, while Plevoets and Van Cleempoel [7] regard themselves as problem-solvers whose very well-informed survey of the subject provides the motivation for the organisation.

Possibly the most romantic in this collection is *On Altering Architecture*, by Fred Scott, published by Routledge in 2008. Scott had already an established reputation for bringing a radical and intellectual approach to his teaching of interiors as the leader of the programme at Kingston University, and rumours of this book circulated long before publication—so it was eagerly awaited [4] (p. xiv). Scott locates adaptive reuse within a wider cultural framework; he places the subject with art conservation, the search for authenticity, the nature of the copy and the reproduction, the ruins of modernity, and, importantly, he exposed the transgressive nature of remodelling, therefore, moving the subject from beneath the authority of the assured architect towards the more disruptive nature of the designer or artist.

Scott speaks with the authority of long academic experience combined with deep knowledge. He develops a sound theoretical underpinning for the subject, the argument, which is developed over 12 chapters, begins with a call to move away from the scarifying process of conservation towards the progressive, or even transgressive, attitude of adaptation [4] (p. 15). The book continues with discussions of attitudes and practices, and it concludes with resolutions—against pastiche and gratuitous improvement [4] (p. 167). All buildings are *"in an imperfect state"* and, therefore, meaning, he suggests, can be created through the *"play between the new occupation and the original use"* [4].

Adaptive Reuse: Extending Lives of Buildings (Birkhäuser, 2017), by Lilian Wong, is a collection of knowledgeable discussions, it is stylishly produced, and it holds a wide-ranging collection of examples. The book is deliberately provocative, and, although erudite, it is also engagingly angry—angry about the reuse of plunder [35] (It is worth noting that given the colonial connotations connected with the term spolia, in the uncompromising discussion of plunder, Wong is careful to never use that term), about lack of considered care for ancient monuments [35] (p. 90), the exploitative nature of facadism [35] (p. 116), tax incentives as the driver for reuse [35] (p. 55), the dominance of the intervention over the impassive host [35] (p. 174), the false historicism produced by zealous preservation [35] (p. 216), and so on. The architect or designer must negotiate a path, she argues, between Frankenstein-like creations of a self-interested monster [35] (p. 244) and that of the compassionate role of the second violinist—supporting the melody of the host building [35] (p. 246).

At the very beginning of the book is a list of quotations alphabetically classified by their focus. The inclusion of this is both innovative and witty [35] (pp. 13–28), so, Ruskin, Douglas, and ICOMOS are all cited under Repair; Watson, and the British Standards Institution under Addition; and the U.S. Department of the Interior and Eugene Viollet-le-Duc under Restoration. The very early statement that Carlo Scarpa's Castelvecchio is *"timeless"* [35] (p. 6) is questionable. It is possible to contend that although it is, undoubtedly, a work of genius and definitely ground-breaking, adaptive reuse has evolved from Scarpa's contrast and analogy approach to one that pursues the concept of "wholeness" [12] (pp. 183–197).

UnDoing Buildings by Sally Stone is a comprehensive study of adaptive reuse that begins with an overview, travels through a discussion about a developed methodology for adaptation, discusses the influence that peripheral areas such smartness, spatial agency, and conservation possess before concluding with an examination about the detailed understanding that the process develops. The book is comprehensive, thorough, and informed. It collects disparate influences and collates them into an organised and influential argument. The publication makes it clear that the process is intrinsically sustainable; that the three tenets of sustainability are a fundamental part of adaptive reuse. It is built upon the urgent need for densification—unlimited horizontal development is no longer ecologically acceptable, therefore, the built environment must learn how to build in on itself, to become more dense, more compact, and more productive.

Adaptive Reuse of the Built Heritage: Concepts and Cases of an Emerging Discipline (Routledge, 2019), by Bie Plevoets and Koenraad Van Cleempoel, makes a very well-informed argument for the primacy of the discipline, which has become *"increasingly important as an urban, architectural, and conservation strategy"* [7] (p. 1). The survey of the historical background and strategic approaches is encompassing, but, in reality, the authors are romantics. Theirs is a search for authenticity, for the fundamental poetry inherent within the patina of time and place. Traces, tradition, and empathy activate the *"creative moment of transformation"* [7] (p. 99), thus, they contend, the patina of the palimpsest evolves into an essential part of the design methodology. Plevoets and Van Cleempoel manage to combine both systems; the first half of this extremely influential book is composed of five chapters of built discussion, while the second half is 19 case studies.

> *I believe a lot in the revelatory capacity of reading. . . if one is able to interpret the meaning of what has remained engraved, not only does one come to understand when this mark was made and what the motivation behind it was, but one also becomes conscious of how the various events that have left their mark have become layered, how they relate to one another and how, through time, they have set off other events and have woven together our history.*

Giancarlo de Carlo, 1990 [36]

11. Conclusions

It is obvious that adaptive reuse is no longer regarded as a difficult, undesirable approach but has moved to the centre ground of the development of the built environment. By the third decade of the 21st century, it is recognised as inherently sustainable, as a healthy, friendly, and economically beneficial approach to the development of the built environment. Adaptive reuse addresses the zeitgeist of the 21st century—the imperative to provide for the basic needs of everyone without damaging the planet, to stop uncontrolled horizontal development, to embrace different ideas and cultures, and to understand the importance of environmentally sound development. It has become the expected approach rather than the exception, the first thought rather than the last resort.

The issues of memory and anticipation that drove the contextual movement have had a direct influence on the evolution of a theory of adaptive reuse. These have encouraged architects and designers to embrace a pluralistic agenda that encompasses the anticipated needs and aspirations combined with an understanding of place. The architect reads the qualities of the building and hears what it has to say.

This emphasis upon the distinct qualities of adaptive reuse has coincided with the rise of diversity in architecture—a subject that has evolved far from le Corbusier's aphorism *"masterly, correct, and magnificent play of masses brought together in light"* but now encompasses much greater scope. The diverse foundations of the subject have allowed an expansive attitude that is more inclusive, embraces difference, is sustainable, but is also creative, technologically advanced, and radical. There is now an expectation that the subject is taught in schools of architecture. Adaptive reuse provides a tangible reality in a world that is increasingly distorted by digital interactions and the rise of AI.

The quantity of these writings reflects the position of the subject within the building industry, and beyond that into wider cultural society, and, as the discussions about adaptive reuse have matured, so the scope is moving. It is becoming less Western-centric, technology is developing, and sustainability in all its forms is directly influencing the evolution of the subject. However, as the discussion about the subject has evolved, so these distinct, pluralistic influences have remained. The contextual base for adaptive reuse, combined with an understanding of the needs and aspirations of the users, has proved to be the starting point for discussion and design.

Funding: This research has received no external funding.

Conflicts of Interest: The authors declare no conflict of interest.

References

1. Benjamin, W. *Selected Writings: Volume 1 1913–1926*; Bullock, M., Jennings, M.W., Eds.; The Belknap Press of Harvard University: Cambridge, MA, USA, 1996.
2. Crouch, D.P. *History of Architecture: Stonehenge to Skyscrapers*; McGraw Hill: New York, NY, USA, 1984.
3. Brooker, G.; Stone, S. *Re-Readings: The Principles of Interior Architecture and the Re-Use of Existing Buildings*; RIBA Enterprises Ltd.: London, UK, 2004.
4. Scott, F. *On Altering Architecture*; Routledge: Oxfordshire, UK, 2008.
5. Robert, P. *Adaptations: New Uses for Old Buildings*; Princeton Architectural Press: Princeton, NJ, USA, 1989.
6. Lanz, F.; Pendlebury, J. Adaptive Reuse: A Critical Review. *J. Archit.* **2022**, *27*, 441–462. [CrossRef]
7. Plevoets, B.; Van Cleempoel, K. *Adaptive Reuse of the Built Heritage*; Routledge: Oxfordshire, UK, 2019.
8. Cramer, J.; Breitling, S. *Architecture in Existing Fabric: Planning, Design Building*; Birkäuser: Basel, Switzerland, 2007.
9. Douglas, J. *Building Adaptation*; Routledge: Oxfordshire, UK, 2006.
10. ICOMOS. Available online: https://www.icomos.org/images/DOCUMENTS/Charters/ICOMOS_NZ_Charter_2010_FINAL_11_Oct_2010.pdf (accessed on 1 August 2023).
11. BURRA Charter. Available online: https://australia.icomos.org/wp-content/uploads/The-Burra-Charter-2013-Adopted-31.10.2013.pdf (accessed on 1 August 2023).
12. Stone, S. *UnDoing Buildings: Adaptive Reuse and Cultural Memory*; Routledge: Oxfordshire, UK, 2020.
13. Jager, F.P. *Old and New*; Birkhäuser Architecture: Basel, Switzerland, 2010.
14. Rodolfo Machado, R. *Old Buildings as Palimpsest*; Progressive Architecture: White Bear Lake, MN, USA, 1976.
15. Cantacuzino, S. *Re/Architecture*; Thames and Hudson: London, UK, 1989.

16. Powell, K. *Architecture Reborn: The Conversion and Reconstruction of Old Buildings*; Laurence King: London, UK, 1999.
17. Jacobs, J. *The Death and Life of Great American Cities*; Random House: New York, NY, USA, 1961.
18. Rowe, C.; Koetter, F. *Collage City*; MIT: Cambridge, MA, USA, 1984.
19. Rossi, A. *The Architecture of the City*; MIT Press: Cambridge, MA, USA, 1982.
20. Venturi, R.; Scott Brown, D. *Complexity and Contradiction in Architecture*; The Architectural Press: London, UK, 1977.
21. Graves, M. Roma Interrotta. *Archit. Des.* **1979**, *49*.
22. Thomas, L. Schumacher, Contextualism: Urban Ideals and Deformations. *Cassa Bella* **1971**, *359*, 79–86, Reproduced in Nesbitt, K. *Theorizing A New Agenda for Architecture An Anthology of Architectural Theory 1965–1995*; Princeton Architectural Press: New York, NY, USA, 1996; pp. 294–307.
23. Murphy, R. *Querini Stampalia Foundation, Venice 1961–1963*; Phaidon Press Ltd.: London, UK, 1993.
24. Jencks, C. *The Language of Post-Modern Architecture*; Academy Editions: London, UK, 1977.
25. 1996 European Union Prize for Contemporary Architecture: Mies van der Rohe Award. Available online: https://www.miesarch.com/work/2752#:~:text=Group%2091$'$s%20Temple%20Bar,city%20or%20town%20in%20Ireland (accessed on 1 August 2023).
26. Matta Clark, G. Serpentine. Available online: https://www.serpentinegalleries.org/whats-on/gordon-matta-clark/ (accessed on 1 August 2023).
27. Jenkins, K. *Re-Thinking History London*; Routledge: Oxfordshire, UK, 1991.
28. Jäger, F.P. *Old & New—Design Manual for Revitalizing Existing Buildings*; Birkhäuser: Basel, Switzerland, 2010.
29. Schittich, C. (Ed.) *Building in the Existing Fabric: Refurbishment, Extensions, New Design*; Walter de Gruyter: Berlin, Germany, 2003.
30. Littlefield, D.; Lewis, S. *Architectural Voices: Listening to Old Buildings*; Wiley: Hoboken, NJ, USA, 2007.
31. Brooker, G.; Stone, S. *Basics Interior Architecture 01: Form and Structure: The Organisation of Interior Space*; Bloomsbury: London, UK, 2007.
32. Brooker, G.; Stone, S. *Basics Interior Architecture 02: Context & Environment*; Bloomsbury: London, UK, 2008.
33. Brooker, G.; Stone, S. *Basics Interior Architecture 04: Elements & Objects*; Bloomsbury: London, UK, 2009.
34. Attiwill, S. Working Space: Interiors as Provisional Compositions. In Proceedings of the Conference Held at the University of Brighton, Brighton, UK, 2–4 July 2009; RMIT University: Melbourne, Australia, 2009. Available online: http://arts.brighton.ac.uk/__data/assets/pdf_file/0018/44811/Suzie-Attiwill_Working-Space.pdf (accessed on 4 August 2023).
35. Wong, L. *Adaptive Reuse: Extending Lives of Buildings*; Birkhäuser: Basel, Switzerland, 2017.
36. Zuchi, B. *Giancarlo De Carlo*; Butterworth Architecture: Oxford, UK, 1992.

Disclaimer/Publisher's Note: The statements, opinions and data contained in all publications are solely those of the individual author(s) and contributor(s) and not of MDPI and/or the editor(s). MDPI and/or the editor(s) disclaim responsibility for any injury to people or property resulting from any ideas, methods, instructions or products referred to in the content.

Article

Assessing the Social Values of Built Heritage: Participatory Methods as Ways of Knowing

Elizabeth Robson 

Division of History, Heritage and Politics, University of Stirling, Stirling FK9 4LA, UK; e.m.robson@stir.ac.uk

Abstract: This paper explores the role participatory methods play in understanding the social values of built heritage, including people's sense of identity, belonging, and place. It is based on research in Scotland where, as in many other countries, there is an increasing emphasis on contemporary significance and public participation within domestic heritage management frameworks. The paper draws on the experiences and findings of a social values assessment for Cables Wynd House, a Brutalist block of flats in Edinburgh that was listed in 2017. Through the case study assessment, conducted over six months in 2019, Cables Wynd House is manifested as a multiplicity of connected realities, diverse experiences, and micro-locations. The participatory methods reveal interactions and tensions between the architectural design and aesthetics of the building and participants' lived experiences and connections. The article argues that the mix of participatory methods provide different opportunities and ways of knowing, surfacing diversity, dissonance, and complexity. It highlights that participatory research is a collaborative process, requiring a flexible and responsive approach to methods. The paper concludes that participatory methods and collaborative approaches can provide nuanced and contextualised understandings of the social value of built heritage, which can complement but also diverge significantly from professional assessments of value. Wider adoption of these methods and the resulting understandings into the management and conservation of built heritage would support more people-centred, inclusive, and socially relevant forms of practice.

Keywords: social value; built heritage; participatory methods; listed buildings; social housing; heritage management; Scotland

Citation: Robson, E. Assessing the Social Values of Built Heritage: Participatory Methods as Ways of Knowing. *Architecture* **2023**, *3*, 428–445. https://doi.org/10.3390/architecture3030023

Academic Editor: Johnathan Djabarouti

Received: 30 June 2023
Revised: 4 August 2023
Accepted: 7 August 2023
Published: 10 August 2023

Copyright: © 2023 by the author. Licensee MDPI, Basel, Switzerland. This article is an open access article distributed under the terms and conditions of the Creative Commons Attribution (CC BY) license (https://creativecommons.org/licenses/by/4.0/).

1. Introduction

Value is a central concept in discussions of heritage and conservation practice. From international conventions to local conservation policies, complex, contextual and, at times, contested values are at play. The values being privileged may be explicit or they may be hidden within 'objective' or professional evaluations of significance. The importance of social values, "the significance of the historic environment to contemporary communities, including people's sense of identity, belonging, and place" [1] (p. 21) is increasingly recognised. However, there are tensions between conserving and preserving while changing as little as possible, the principles that have traditionally guided heritage and conservation practice [2] (p. 1), and maintaining a place's significance to communities in the present. These challenges have meant that "*in practice* historic and aesthetic values tend to override others, such as social value, in heritage significance assessment" [3] (p. 6, italics original).

A focus on significance assessment as the first step in heritage management has become accepted practice in national and international conventions [4] (p. 9). The Burra Charter [5], issued by ICOMOS Australia in 1979, has been widely credited as a key development in shaping these practices [6]. The initial Charter attempted to bridge the divide between tangible and intangible heritage with a range of (theoretically equal) types of value considered as part of establishing significance [1] (p. 23). It states, "Cultural significance means aesthetic, historic, scientific or social value for past, present or future generations" [5] (Article 1). As Lesh explains [7], the inclusion of social value in this definition owes a lot to the specific history of conservation in Australia and what is meant by the term has varied in practice over time. Subsequent revisions of the Charter have given greater prominence to the social values of contemporary communities as an important part of cultural heritage [8,9], seeking to clarify what has been an evolving concept and reflecting the influential people-centred processes adopted by Australian heritage practitioners since the 1990s [7] (pp. 55–56). Although initially emerging in a national heritage context, these ideas have been developed through international heritage instruments such as the 2005 Council of Europe Framework Convention on the Value of Cultural Heritage for Society [10] and have been widely adopted in other national contexts. For example, the influence of the Burra Charter can be seen in English Heritage's Conservation Principles [11] and it is expressly cited in the definition of cultural significance given in Historic Environment Scotland's national policy [12]. This growing emphasis on social values has raised the prospect, at least in theory, of broader and more inclusive conservation frameworks and practices.

Discourses on the value of heritage have ancient roots and, in a Western European philosophical tradition, are often based on well-established ideological positions that characterise the purpose and power of the arts more broadly as: innately beneficial; making a positive contribution; or having a negative impact [13]. The perceived dichotomy between the intrinsic and instrumental effects of heritage (mirroring the first and second of these positions) arguably obscures a more nuanced discussion on the complex interplay between co-existing values. The nature and role of values in heritage policy and practice have been interrogated by a growing body of critical heritage scholarship. Emerging alongside the international and national developments in heritage policy described above, critical heritage studies have foregrounded the contextual, relational nature of heritage [14,15], questioning how heritage is identified, legitimised, and mobilised [16–18]. The idea that values are inherent in the fabric of a building or other material, a principle that characterises many conservation instruments (including the Burra Charter), has been critiqued by critical heritage studies scholars, who highlight that heritage is embedded in, and a product of, social and political processes [19–21]. These debates have led to a more nuanced response to the 'things' of heritage that integrates different aspects of significance [15,22,23]. Scholars of critical heritage studies have also explored how meaning and values are formed and expressed, emphasising that values are fluid and dynamic expressions of continuous processes of valuing [14] (pp. 45–46), [18] (p. 167), not definitive or singular but plural and liable to change or evolve in response to the wider context and the practical "performances" [19] (p. 3) associated with their (re)generation. Heritage professionals and conservation practices are identified as active participants within these on-going processes of negotiation and interaction [1,18,24]. These new perspectives on value and the processes of valuing have been raised alongside more practical questions, such as what a "values-based approach to culture resource management" might look like [25] (p. 89) and what "the validation of multiple conceptions of value" might mean for conservation practices that are rooted in "processes which involve the fixing of meaning and value" [26] (p. 1).

Values not only determine what is prioritised or conserved as heritage and how, but also who gets to participate in those processes. In common with many other countries, community participation and social values are increasingly prominent in the heritage policies of the UK nations (the built environment being a devolved area of public policy), but the implications have been slower to filter through into day-to-day practice. Despite some progress in recent years, heritage is principally "a field in which specialist practitioners and decision-makers consult with local people and (sometimes) facilitate their involvement" [27] (p. 3), [4,6,28]. There are exceptions but, in many situations, communities are talked to, rather than listened to, about the significance of places that they are familiar with and value [29] (p. 141). Scholarship from critical heritage studies has highlighted the contradiction between policies of increased community participation and the established reliance in practice on expert judgements and professional authority [3] (p. 67), [30] (p. 51), [19,31,32], critiquing heritage and conservation practice as reproducing dominant power and knowledge hierarchies, what Smith terms the "authorised heritage discourse" [19]. However, there has also been recognition that responding to value as multiple, dynamic, and contextual presents practical and theoretical challenges for practitioners working within institutions and systems that are based on established principles for conservation practice [1,26]. A key problem that has been identified within both critical heritage studies and international heritage management debates is the absence of appropriate methods for the assessment of social value in 'real world' contexts [3] (p. 67), [4] (p. 28), [14], [24] (p. 69), [33], [34] (pp. 145–146, Points 1, 2 and 4).

Without appropriate means for practitioners to understand and evidence the social values of the historic environment they remain invisible in official assessments of significance. Avrami et al. note that, "[i]n order for conservation planning processes to center on, and take into deeper consideration, the multitude of social values, we need to develop better tools and methods for the assessment of cultural significance" [24] (p. 69). With the notable exceptions of the edited volumes from Sørensen and Carman [35] and, more recently, Madgin and Lesh [36], methods have received relatively little attention to date from within critical heritage studies. Nonetheless, alternative, qualitative methods have gradually emerged, especially from countries with significant indigenous populations, such as Australia [1] (p. 28), [3] (p. 5), [33,37], [38] (p. 571). Academics have successfully applied collaborative approaches that centre community knowledge in a variety of heritage studies contexts [39–43]. These studies have established the strengths of using participatory methods in research with contemporary communities and have identified them as "fruitful avenues" for exploring social values [4] (p. 34), but they have not explored how these different methods enact and (re)produce different knowledges.

This paper addresses the current gap in the academic literature on the 'work' that methods do in values assessments and how they operate in context. In keeping with critical heritage studies scholarship, in this study values are understood as dynamic and plural, contextualised expressions of on-going social processes of valuing. The study draws on the experiences and findings of a social values assessment for Cables Wynd House, a 1960s Brutalist block of flats in Edinburgh, Scotland, which is still occupied and in use according to its original design. Cables Wynd House was selected as a case study within my doctoral research [44] because the assessment was expected to provide critical insights on the implementation of rapid, participatory methods in complex social and environmental contexts. In addition, Cables Wynd House was listed as a nationally significant building in 2017. In Scotland, buildings are listed based on an assessment of their "special architectural or historic interest" [45], a continuation of earlier policies that reflect the parameters of the relevant legislation [46]. Having been through a formal listing process relatively recently, the case study offered the scope to explore whether that process, or the building's listed status, has contributed to the social values and the extent to which the conservation priorities identified in the listing are congruent with the values expressed by residents and other communities.

Through the Cables Wynd House material, I explore how a range of rapid, participatory methods provide different opportunities and ways of knowing. Looking at the assessment findings, I show that participatory methods and collaborative approaches can provide nuanced and contextualised understandings of the social value of built heritage, surfacing diversity, dissonance, and complexity. The variety of communities and values identified and the ways in which people experience Cables Wynd House, as a place and as a focus for heritage conservation, also highlight how social values can diverge significantly from professional assessments of value. By looking at the methods comparatively, and how they operate in combination, I demonstrate that method choices are not merely neutral, technical decisions; they (re)produce different types of knowledge and actively shape the resulting understandings of value. This is an important insight, of significance for academic research on heritage values and for conservation and heritage practice.

In the following sections, I first introduce the Cables Wynd House case study and the methods that were adopted in the social value assessment. This is followed by a description of the results of the assessment. I then comparatively discuss the understandings achieved through the different methods applied in the study and how the assessment approach was adapted to the specific context, demonstrating the flexible and responsive approach to methods that this type of research requires. The paper concludes with some reflections on the practicalities of applying multi-method participatory approaches to understand the complex and dynamic social values of built heritage, arguing that wider adoption of these approaches, and incorporation of the resulting knowledge within conservation and heritage management, has the potential to generate more inclusive and socially relevant forms of practice.

2. Case Study Site and Methods

Cables Wynd House, also known as the Banana Flats on account of its distinctive bend (Figure 1), is located in the Kirkgate area of Leith, Edinburgh. It is embedded in a complex and dynamic urban context, both socially and environmentally. Leith is a part of Edinburgh that has been shaped historically by the presence of the docks, port, and industrial manufacturing. Today Leith is a culturally diverse area of housing and light industry, with good transport links to the centre of Edinburgh. According to the 2011 census, significant percentages of the population were born outside the UK (19.6%) and self-identify with ethnicities other than the majority White Scottish or British, including Polish (11.4%) and Asian, Asian Scottish, or Asian British (5.3%) [47] (Output Area S00107051). Leith also encompasses some of Scotland's most deprived areas when it comes to indicators for income, employment, education, health, access to services, crime, and housing. The Scottish Government's Index of Multiple Deprivation indicates that the data zone including Cables Wynd House is within the 5% most deprived data zones in Scotland [48] (Data Zone S01008788).

Figure 1. Exterior of Cables Wynd House—East-facing/private balcony side (photo by the author).

The Kirkgate development that includes Cables Wynd House was one of several large public housing developments that were constructed in Leith between 1963 and 1965 [49] (p. 367) following slum clearance programmes. The House contains 212 flats, laid out over ten stories, accessed via communal landings and with private balconies to the rear.

It is owned and managed by Edinburgh City Council, the residents being mainly Council tenants with a small number of owner-occupied flats. In January 2017, following a period of public consultation, Cables Wynd House was added to the national list of buildings of special architectural or historic interest at category A: outstanding [50]. The Statement of Special Interest cited reasons related to the design, which is in the New Brutalism style and reflects the then "emerging theoretical interest in community planning, using external access decks as a way of recreating the civic spirit of traditional tenement streets" [50] (p. 9). As well as its architectural interest, the Statement of Special Interest notes that the building and its location have frequently been used as subjects for photography and filming, and feature in Irvine Welsh's novel, Trainspotting [50] (p. 8).

The social values assessment for Cables Wynd House discussed in this paper was conducted principally over a period of six months (March–September 2019) and adopted a rapid, participatory approach, applying a mixture of qualitative methods that are principally drawn from ethnographic practice. The study began with observational techniques, including behaviour mapping, drawing a rough plan of the location, and recording what was observed, the behaviours that were displayed at different places, and how people moved around the site [37] (p. 90). Observation was also conducted during subsequent on-site activities, which were carried out during repeated daily visits (13 in total). Many ethnographers emphasise the importance of shared practice in producing understanding [51], but Davies argues that the "nature, circumstances and quality of the observation" is also key [52] (p. 83). Such attentive observation not only sensitises the researcher to the social and environmental context, but can also help in identifying areas for further enquiry through other methods and suggest future research directions [53] (p. 47). In this case, the in-person and on-site activities were supported by online observation of public participatory media (Instagram and Facebook) posts related to Cables Wynd House and a review of the documentation and publicity surrounding the listing process in January 2017.

The study applied a combination of participatory and co-creative research activities. The approach taken was to engage with local authorities and community groups as an initial point of contact and to reach out to local residents and tenants through them. Semi-structured interviews (seven in total) were conducted early in the research period with heritage practitioners, Council officials, and representatives of local organisations (four respondents) and then later in the research period with tenants (three respondents). All semi-structured interviews were conducted one-to-one and in-person, and one incorporated a walk through the area surrounding the House. The semi-structured format was chosen to allow for a relatively free-flowing discussion and mutually engaged exchange from which a depth of understanding can be achieved [52] (pp. 113–115). Given the large number of people living at Cables Wynd House, a structured interview technique was also proposed. Following the approach described by Taplin et al. [37] (pp. 87–88), a short, six question interview format was developed, to be conducted quickly, either face-to-face or by self-completion. The emphasis was on open-ended questions, to provide scope for qualitative responses. A folded A4 leaflet containing basic project information and the six questions was distributed to all the flats in the House (two responses received). The same questions were used subsequently for face-to-face interviews with people passing through the public areas of the House and gardens and on the street to the front of the building (eight respondents). Participants in face-to-face interviews were a mixture of tenants, local residents, and visitors.

As well as the individual participatory interview methods, group activities were proposed to explore how the interactions and negotiations between individuals shaped the values being expressed. Although there has been an active residents' association for Cables Wynd House in the recent past, there was not one in place at the time of this study. However, there are several organisations in the immediate area providing community spaces and services, which offered alternative opportunities to connect with existing activities and social groups. Following enquiries at a local community centre, it was possible to trial a photo-elicitation activity [54] (p. 452), working with a group of older people living in the area. This activity took place over three consecutive weeks, each of the sessions lasting one to two hours. The first week was a group meeting at the community centre to introduce the research and agree together how we would arrange the activity. All group members were invited to participate, and five self-selecting volunteers (four female, one male) were willing and able to do so. Some of the participants were more familiar with Cables Wynd House than others, but none were present or former tenants. The following week, the participants met to take photographs (using their own cameras) and I accompanied them. Participants were asked to focus on things that they felt were significant about the building or the wider area. We visited the communal areas of Cables Wynd House as a group and then split up to explore the surrounding area individually or in pairs before reconvening. The third week, the group met at the community centre to share and discuss the images the participants had taken. Each participant selected around 10 of their photographs to share and speak about. Afterwards, with the agreement of the group members, a number of the photos were selected and printed as A3 or A4 colour images and displayed on two of the wall mounted noticeboards in the main entrance vestibule of Cables Wynd House (Figure 2). The photo exhibit was left up for five days with a comments sheet (unfortunately, this was removed during the week) and I spent a couple of hours on site on the first and last day to take comments in-person (20 responses, 18 from tenants and 2 from visitors).

Figure 2. Photo exhibit in the vestibule of Cables Wynd House (photo by the author).

3. Different Communities and Multiple Values

Cables Wynd House (hereafter 'the House') is a place of residence, employment, and (through being home to so many people) a social hub. Council staff advised that the flats are normally almost all occupied, although there has been an increased frequency in turnover (potentially related to changes in policy for managing housing stock). The research identified a number of communities of interest, identity, and location, for whom the House is of significance. These included: current and past tenants; their friends and relations; people "born and bred" in Leith and/or identifying as "Leithers"; and younger people, including those making use of the park and basketball court situated to the rear of the building. There are some communities for whom the House is principally of significance for reasons closely aligned to the listing criteria (i.e., those with a professional or personal interest in architecture and design or the literary/film connections). This was reflected in comments from some respondents and images and comments posted online. Design and aesthetic factors were also mentioned by some participants as supporting other social values, as is reflected in the discussion below.

First and foremost, the House is experienced as a home and, for the most part, a place of safety and belonging. Since it was constructed in the 1960s, it has been a place of residence for many hundreds of people. Some tenants move into the House after being in temporary accommodation or homeless. One man who responded to the questionnaire indicated that his strongest memories were of his first night staying in the House: "Having been homeless for some time it was a great relief despite having no furniture at all" (Respondent 3.22). Although there may be a tendency to think of a 20th-century concrete building as modern and relatively recent, there has been more than enough time for three generations to have grown up living in or around the House. During the study, various multi-generational or family connections were mentioned. For long term residents, there were memories of their own childhoods and bringing up their children in the House. There were inter-generational connections involving non-residents, as in the case of one young woman, who said, "My aunt's lived here over 20 years, [I've] visited regularly all my life" (Respondent 3.28). There were also instances of multiple generations living separately in the House; for example, one resident of 20 years indicated that, "My daughter stays here too" (Respondent 3.30). One resident viewing the photo exhibit said, "Nan grew up here, me, my mother-in-law, a lot of history" (Respondent 3.49). In other contexts, three generations of association might be expected to lead to memories, attachment, and value, but this multi-generational connection seems to have passed largely unremarked upon in the social housing literature, perhaps because of the individual nature of tenancies. However, in this case, family history was important in how the House was viewed and valued.

In addition to the building operating in its primary function as housing, it is also a hub for, and generative of, numerous social networks, relationships, and interactions between tenants, local residents, staff, and visitors. These relationships and the sense of community are central to how the House is valued and, for those with an active social network, support feelings of safety and belonging. However, as participants reflected, these values and experiences varied between people and over time. One interviewee indicated that "everyone knows everyone", but reflected later that, "people in here find it quite lonely" (Respondent 3.7). Others felt the "community spirit" had declined since their early years in the House, as their children grew up and tenants changed, with a sense that today people are "not encouraged to try and meet and talk" (Respondent 3.5). This suggests that being connected to "everyone" is limited to within particular social groupings or contexts. It was also apparent that the physical space itself shapes these social interactions. Participants spoke about talking to people on or from balconies, but also mentioned the absence of social opportunities or physical spaces for interaction. Visitors also commented on the fact they did not see many people in the communal areas when they were in the House: "Actually, when we were in [the flats], was sort of like a ghost town" (Respondent 3.10).

As the above suggests, the feelings expressed towards the House were complex and mixed, at times strongly expressed and at other times more equivocal. The day-to-day experiences of living in the House prompted nostalgic memories of time spent with family members alongside frustration, anger, or resignation over the management of the property and realities of living in close proximity to other people. Participants mentioned disruptive works in the kitchens, mice getting into the flats, broken heating, and noise from neighbours or the basketball court. A few people had experienced violence or disturbances that had left them feeling unsafe in the House. As in any (particularly high density) residential area, it is the behaviours of their immediate neighbours that impact most directly on residents. Tenants tended to characterise the House according to their landing; for example, "never had it [drug dealing] on my section of the landing—quietest bit" (Respondent 3.5). Such comments demonstrate a hyper-local understanding of the House and how people identified with places through individual relationships and shared behaviours. Similar spatial differentiation within what might otherwise be seen as a unitary area of housing is described by Pendlebury et al. in their study of the Byker housing development in Newcastle [55] (p. 188).

This hyper-local reading of the House itself sits alongside attachments and identities expressed in terms of the wider area. The communal landings in the House provide a wide view of the local surroundings and almost all of the photo-group images prompted discussions on points of reference within the wider landscape and how it had changed, such as: "there used to be the big store there and there was shops and then the wee [small] park" (Respondent 3.10). Participants were often explicit about being from or wanting to live in Leith, with comments such as: "Lived in New Town [another area of Edinburgh] ten years and here five years. Wish I had spent that time in Leith" (Respondent 3.6); "Born and bred in Leith... wouldn't want to be anywhere else" (Respondent 3.27). These expressions of identity, belonging, and attachment contrast with feelings of aversion and examples of participants distancing themselves from the area or the House due to what one participant referred to as "the Trainspotting stereotype" (Respondent 3.29), a reference to the drug addiction and economic depression described in Irvine Welsh's book and the film of the same name. Sometimes these contrasting feelings were expressed by the same respondents at different times, depending on the specific experiences, places, or identities being foregrounded in the discussion. For example: "I was here a lot while young [...] hasn't changed a lot, still a nice place to be"; and then later, "People [are] stuck here forever, never going to get out. People came here with young children, want to get out and have a garden. Kids can't get out and play, disabled kids, can't keep an eye on them when up high. [I] have a young baby now and eventually will get out" (Respondent 3.7).

The degree to which the House is perceived to be receiving care, whether from the Council or the tenants, affects how it is valued. While comments were principally concerned with the physical appearance of the House, they reveal feelings that go beyond the present physicality to reflect lived experiences and social relations, past and present. For example, the issue of cleanliness was referred to when discussing the decline in "community spirit": "Tenants used to clean the building, wash the landing every day for a week and then passed on to the fifth flat and on like that, and clean the stairs between the landings once a month" (Respondent 3.5). Although experienced in the context of a specific building or neighbourhood, studies have shown that changes in social cohesion and neighbourly reciprocity of the sort described are society-wide issues [56,57]. The links being made between cleanliness and behaviour suggest that "clean" and "cared for" are experienced and understood not only as material matters but as "analogies for expressing a general view of the social order" [58] (p. 4). Perceptions of the building are therefore influenced in part by how respondents are positioned with regard to the social structures and behaviours that they associate with the House.

There was a high level of awareness among people contacted as part of the study that the House had been listed (90% of structured interview participants indicated they were aware of the process), with responses ranging from interest to incredulity. While most indicated that the formal status had not changed their feelings towards the site, building on the point above, some people expressed a disconnect between the interest taken in the building and the attention paid to residents' interests and priorities: "The Queen has a listed building, but does she have problems with heating like us?" (Respondent 3.7). Some respondents also associated the listing with a perceived lack of maintenance: "Since the new status it has gone downhill drastically. Council are hanging back [on repairs]" (Respondent 3.5). However, for some tenants, the design features supported their attachment to the House: "From the outside the first impression is not so good but when [you] go inside and see the design of the flats, if into design, then you change your mind" (Respondent 3.6). This same respondent referred to the building as "iconic" and said, "[I] feel privileged, fortunate to be in it", while several people responding to the photo exhibit described the House as "unique". Against the backdrop of a proposed listing, Pendlebury et al. [55] explore how the neighbourhood of Byker in Newcastle is valued by residents and professionals (what makes it "unique and special") and the potential impact of the listing. They conclude, as seen here, that the listing itself was not an especially important issue, barring concerns about future improvements to the building or marginal benefits (p. 197).

Communities are collective and therefore relational. While a simplified community identity may be presented externally, membership is more accurately a complex and evolving negotiation [59] (p. 21–22), [60] (p. 132). The contextualised nature of community membership and identity was apparent in this case, with the residents of Cables Wynd House manifesting as a heterogeneous group and the House as multiple micro-locations. Experiences were highly differentiated between groups and individuals, impacting on how people valued the House and felt connected to specific places. The lack of a functioning residents' association is also suggestive of a degree of fragmentation within the wider tenant community. Participant responses indicated an awareness of this diversity and dissonance, but the degree to which they aligned themselves with groups, values, and behaviours depended on contextualised identities and experiences. As an attribute, "community spirit" depended largely on personal experiences and networks of active relationships. Some people felt there was a greater sense of community in the past, for others it was a positive aspect of their current experience or thought to be improving; yet it was evident from other comments that in practice some people and groups may be isolated or excluded. This isolation could be physical or embedded within concepts of community and place. For example, values of community belonging that emphasise being "born and bred" locally could operate to exclude other communities and experiences in an area known to have experienced significant in-migration.

Exploring the range of experiences and views of the physical, social, and emotional environment of Cables Wynd House, including seemingly incompatible or opposing values and practices, was important in revealing how values and understandings of place were operating within the particular context. Although the House has a much smaller body of residents than the Byker housing development, the observations of differentiated, complex, and contradictory values in this case mirror the findings of Pendlebury et al. in their study [55]. Also taking Byker as one of his cases, Malpass [61] expands on some of the challenges inherent in taking occupied "council housing" and valuing it according to formal listing criteria. One of the critiques he identifies is that, "listing tends to place heavy emphasis on the building itself, as an object of importance in itself, abstracted from the context in which it was created and separated from the people who use and interact with it" (p. 205). As was found in this case, improvements to the physical environment (whether for conservation purposes or to upgrade living conditions) may not be experienced as care and attention if lived experiences more broadly are of social disruption and disregard. It follows that preserving the social values of home and belonging (and the "civic spirit", identified in the listing document for Cables Wynd House as one of the design ambitions of the building) depends on more than maintaining the structure. It requires an understanding of and support to the social processes associated with the building and the communities that call it home [61].

4. Discussion

4.1. Methods as Ways of Knowing

The various methods used in the Cables Wynd House assessment engaged different groups and enacted different sorts of knowledge, resulting in multiple and diverse under-standings of the House. Observation (on site) and behaviour mapping helped to build an understanding of the spatial context. The mapping was not limited to the public areas of the House but extended to include the surrounding area, the locations of local services, where people gather, and the routes taken between areas. This revealed how people actively engage with and construct the landscape through practices that do not necessarily follow planned uses of the space [62]. Not all practices or interactions can be readily observed, either because of what they are (such as unsanctioned practices) or when and how they take place (for example, gatherings in people's homes). As one woman's response to the photo exhibit indicated, time of day and weather can also impact on behaviours: "[The photo exhibit has] not captured the true meaning of the flats. I've been here two years, Friday or Saturday night or a sunny day and that park will be full of people drinking, I can hear and see them from my balcony" (Respondent 3.44). Differences between observation and the understandings gained by other methods can also suggest areas for further investigation [53] (p. 47). For example, during visits to the House, I regularly observed people working on the volunteer gardening project and cleaning or maintaining communal areas, but interview responses reflected a perceived lack of care and attention to the House, an area of dissonance that was brought into particular focus through the responses to the photo exhibit. Similarly, in-person observations can evidence activities or communities that may be absent from the discussions or choose not to engage in the research activities.

Structured and semi-structured interview methods were critical to making sense of the House and exploring or understanding observed activities. After securing permission to access the communal areas of the House, discussions with City Council and Historic Environment Scotland staff familiar with the management arrangements and listing process provided useful contextual background for the study. The subsequent interviews with tenants and local residents highlighted the detailed and distinct knowledge held within communities. Although interviews were based around prepared questions, during the discussion respondents frequently developed new ideas or suggested new avenues for enquiry. On occasion, respondents shared experiences and memories that related to past or concurrent identities at different times in the interview. This resulted in seemingly contradictory statements, which were not resolved or clarified relative to one another (each being consistent with that part of their story), but which emphasised the dynamic multiplicity of values associated with the House. By drawing on the material from multiple interviews, it was possible to gain an understanding of the range of experiences encompassed within or across communities.

The self-completion rates for the questionnaire posted to the flats were very low, with less than 1% of leaflets returned. However, the two responses received each reflected very different experiences of residency; one highlighting feelings of safety after being homeless for some time, the other recalling disturbances and "being woken at 4am by a junkie looking for a fix" (Respondent 3.21). Edinburgh City Council's current strategy is to prioritise homeless tenants for housing in Cables Wynd House, but the questionnaire response was the only time that someone self-identified as formerly homeless or specifically mentioned that past experience. This suggests that the opportunity to respond anonymously provided a safe space to share this memory. As was anticipated, the in-person approaches for structured interviews had a much higher response rate [63]. The experience of being on site and speaking to people directly also contributed to my understanding of who was present, how they were using the space, and why. Negative responses to in-person requests for participation were themselves revealing, identifying absences and potential challenges to engaging people in other research activities. Information from semi-structured interviews with staff from the City Council suggested a relatively small number of tenants were non-English speaking. However, my inability to engage with non-English speakers was a limiting factor in over 10% of the structured interview requests made in-person. This resulted in some recognised gaps in participation, which were highlighted in the assessment report. Acknowledging the inevitably partial nature of the assessment findings was important, as focusing only on what is known risks reinforcing existing gaps and silences. Identifying at least some of the realities that are excluded also underlines the open-ended and contingent nature of values assessment, an appreciation for which could be critical when considering the findings as part of future management actions.

The photo-elicitation activity engaged participants with the multi-sensorial aspects of being in place and moving through the area [64] as they captured images that conveyed aspects of the House or surrounding area that were of significance to them. The group members' engagement resulted in a selection of 54 images that were then reviewed together.

The existing familiarity between group members and our shared experiences of visiting the House resulted in a free-flowing discussion, during which people made observations about the location, how places had changed, and personal or family connections to the area. Participants who had initially indicated that they had only passing familiarity with the House shared detailed knowledge of the area and, in some cases, of the House as well. Respondents not only spoke about what was in the pictures, but also things that were not visible, past experiences, and absences. The exchanges between group members demonstrated how interaction and negotiation between individuals shapes the values being expressed and were also revealing of different values or associations. Two similar images of a communal corridor were described as showing variously:

- "the pride they [the tenants] took in their area, bright coloured doors, no rubbish to go out" (Respondent 3.11).
- "the sameness and the similarity [of matching doors down the corridor]" (Respondent 3.9).

Viewing one another's photos began to "break the frame" [65] (pp. 20–21) of taken for granted views, opening up more reflective discussion. For example, this exchange in response to a photo of the communal balcony with the sunlight coming in through the windows onto a shiny floor (Figure 3):

Figure 3. Cables Wynd House communal landing (photo by Respondent 3.8, reproduced with permission).

Respondent 3.9: Huh, that makes it look. . . quite attractive.
Respondent 3.10: It does!
[laughter]
Respondent 3.8: You can see how clean it is.
Respondent 3.12: Mmhmm.

Respondent 3.10: As I say, it's a few years since I was there, but then it was all graffiti and horrible.

Respondent 3.12: That's remarkable actually, the whole building, there was not one bit of graffiti that I could see, in the whole building.

The subsequent photo exhibit in the communal vestibule of the House further enhanced the depth and range of engagement, revealing important areas of dissonance. As in discussions with the group members, people were observed identifying places they knew and there was some evidence of people having physically touched or drawn on the images (though this did not happen while I was present), illustrating how a physical artifact or image can be used to prompt interaction and reflection. During the group discussion, the photographer was able to explain the intention behind their image and the experience connected to it, constructing a narrative beyond what the picture showed [66]. When the images were displayed in the photo exhibit, they were left completely open to interpretation and were used by respondents to "produce and represent their knowledge, self-identities, experiences and emotions" [67] (p. 82). Responses to how the building was shown in the images, and perhaps also the range of pictures taken and selected for the exhibit (by non-tenants), identified a disconnect between how the House looked compared to how it was experienced. For example, several people responded with comments such as: "[It] looks a lot different to how it looks when you're in it [. . .] looks very clean" (Respondent 3.43). These comments resonated with interview responses, in which people had focused on cleanliness not only as a physical or practical concern but as an expression of "community spirit" in the House.

Combining multiple methods not only supports a greater depth of understanding but can also inform the emergent research process. The combination of structured and semi-structured interviews served to identify common touchpoints and supported the interpretation of observations or other activities. The understandings and knowledges each provided usefully complemented one another; the structured interviews suggested potential areas for discussion in the more detailed and in-depth semi-structured interviews, as well as an indication of how widespread the specific experiences and associations mentioned by the semi-structured interview respondents were. Although the response rate to the self-completion questionnaire was low, a couple of people did bring the leaflets to the semi-structured interviews, showing that they had helped raise awareness among residents that the research was taking place. The impact of sequentially implementing methods is also seen more directly in the outputs of the photo-elicitation being taken forward into the photo exhibit. While the sequencing was partly practical, as it took time to identify people willing to participate in activities, starting with more general exposure was helpful in building up familiarity with the site, key individuals, and the wider context, which proved to be important when it came to implementing more engaged methods and interpreting the resulting materials. Towards the end of the assessment process, the draft findings were shared with participants and professionals responsible for the conservation and management of the House. A poster summarising the key findings was also developed, to provide feedback to tenants and visitors who may have observed or participated in the research but not provided contact details. These activities were part of my accountability to the original knowledge holders, but also intended to increase awareness of the diversity of values associated with the House and the different ways in which it is experienced as a place, beyond its observable functions and physical form, providing important social context for any future conservation actions.

4.2. Methods in Context

Working with multiple methods and combining different types of method generated a plurality, or a breadth as well as a depth, of understanding. While at times the participant engagement led to seemingly contradictory or opposing statements, the aim was to "obtain a variety of interpretations rather than seek consistencies" [52] (p. 109). As a result, the assessment built up a complex understanding of the diversity of social values for the range of communities with interests in Cables Wynd House. The "methods assemblage" [68] used in the assessment was able to reveal and accommodate this complexity. More than a cluster of methods, this approach looks beyond methods as techniques to also consider how they are embedded in dominant epistemologies and hierarchies of knowledge, systems for reporting and recording, and complex material, individual, and organisational/community relations [68] (p. 160). These are all factors that affect how methods work in different contexts and when implemented by practitioners with different profiles. In this case, the fact I was working alone on the assessment meant personal attributes, such as gender and language, impacted on engagement. While I was careful to consider my positionality, it was often in the processes of reflection and interpretation that unconscious biases or gaps became apparent. For example, the choice of terminology was flagged in my draft report, where there might be implicit negative connotations in referring to Cables Wynd House as an estate, as opposed to the more neutral development or building. Such reflections served as reminders that, although the report was based on participatory methods and included community voices, the process of analysis, interpretation, and writing inevitably privileges and is shaped by a researcher's "theoretical and epistemological commitments" [69] (p. 12). Working with others (in this case my supervisory team) who have complementary but diverse specialisms can assist in identifying and compensating for unconscious bias, as well as supporting a multi-methods approach [70] (p. 42), [37] (p. 81), [71] (p 352).

Drawing on a range of methods means that the research process can be responsive to dynamic contexts and developing understandings. In anticipation of the complexity of an inner-city context with relatively high levels of diversity and transience, the proposed approach for the Cables Wynd House assessment was to deploy rapid, participatory methods over an extended period. In practice, both the overall amount of time and duration of the study had to be increased to obtain sufficient material for the assessment. This was principally due to challenges in identifying community structures and engaging participants in the methods. Given the social context and the lack of a residents' association for the House, it was expected to be challenging to engage participants, particularly in more collaborative activities or those requiring repeat engagements. Following the low response rates to self-completion questionnaires, plans were adapted and in-person structured interviews were scaled-up. Referrals by formal gatekeepers made it possible to conduct semi-structured interviews with a small number of tenants. As a site embedded in day-to-day life, it was planned to work with residents to develop photo or written diaries of their daily engagements with place. Through notices posted in the communal areas of the House and observation in the wider area, I identified several community organisations holding events for local residents, either at the site or in the immediate vicinity. It was possible to attend some of these community gatherings and they provided opportunities for engagement. However, it remained difficult to identify and engage the informal, personal networks that were described by interview respondents as contributing to their sense of community.

Online observation and mapping the local area gave me an understanding of some of the local amenities. Enquiries about possible collaboration initially received positive responses from two community support organisations in the area, one working with young people and the other with women. Unfortunately, practical constraints, competing priorities, and limited resources meant that the planned research activities could not be arranged with those groups. While visiting a local community hall, I saw an activity advertised for older residents interested in photography. Unlike the other organisations that I had approached, I was able to engage directly with group members and the collaboration did not require support from the co-ordinating organisation or divert scarce resources and time from other priority projects. The proposed activity aligned well with the group's existing activities and the members agreed to focus on Cables Wynd House, resulting in a productive series of exchanges that also generated the material used in the photo exhibit at the House.

Although there were some initial assumptions, based on existing documents, about potential communities and the types of values and practices associated with the House that might be significant, the outcomes of the social values assessment were not predetermined. The assessment was a process of exploration, undertaken together with the participants. Like all socially-engaged research, it required a flexible, responsive, and reflexive mode of practice, working with emergent understandings and evolving contexts. Whether a planned method could be implemented depended on a combination of factors, not least the willingness and availability of potential participants, networks of relationships, and the wider context. As with all participatory research, I had a responsibility to ensure the process was conducted ethically and to take cognisance of the potential impacts that the activities or findings may have on the individuals and communities involved. When assessing social values, there is always the potential for activities to surface potentially distressing or emotional issues, tensions, or conflicts for participants. The memories associated with historic places are not necessarily the "nostalgia of good times past" [72] (p. 58). This understanding, combined with the emergent nature of the research process, meant that the social values assessment demanded an "everyday ethics" [73] (p. 127), a continuous process of self-reflection and dialogue with participants, in which consent is renegotiated and reconfirmed throughout the process and in response to the specific context.

As this study shows, working with methods in context is not a purely technical matter. It implies new ways of thinking about knowledge production and of working with the individuals and communities who are expert in their own relationships to place. The history of current conservation practice has been dominated by Western European thinking and positivist traditions that emphasise scientific processes, with professional judgements presented as objective and constant. Adopting more participatory, responsive, and reflexive methods unsettles this established reliance on professional judgements and changes the role of the practitioner from being the only (or in some cases even the primary) expert and custodian of built heritage. While this shift in power and authority brings challenges, such approaches can result in new, shared understandings of the range of values associated with built heritage, ultimately supporting its future conservation.

5. Conclusions

The Cables Wynd House study is one example of how a multi-methods participatory approach can be applied to explore the variety of communities and range of social values associated with built heritage. There were unexpected challenges and adjustments required throughout the process but, through working flexibly, it was possible to implement a range of participatory methods. The understandings that resulted from the assessment depended on the combination of methods and an iterative, close examination of the resulting material [69]. This depth of knowledge could not have been achieved through non-participatory, desk-based research alone, and differs significantly from professional assessments of value, as detailed in the listing documents. It was apparent that people were aware of the building's status as nationally significant, but the values that they associated with the House were rooted in their day-to-day experiences, relationships, and intimate knowledge of the place over time. That said, there is a relationship between the formal heritage processes and the social values identified. The architectural features that principally underpin the listing do impact on how the House is valued; just as the experiences of the House as a place of home, community, and connection, which are central to how people value the building, are also reflected in the original design intention. The case study also shows how conservation and management actions focused on the physical fabric have the potential to strengthen or undermine these social values, depending on how they resonate with people's other understandings and experiences of place (see also [74]).

Each method provided insights and generated material that, taken together, informed the overall understanding of the social values associated with the House. However, the methods were not simply alternative means of achieving the same understanding, as was apparent, for example, in the differences between the knowledge shared during an individual semi-structured interview when compared to the negotiated understandings and different views that emerged from the photo group discussion. The different methods provided different ways of knowing, with different knowledges negotiated and (re)produced through the process. The multi-methods approach was therefore critical in surfacing and understanding the complexity and range of values associated with the House. Furthermore, the knowledge generated through the different methods did not straightforwardly make up different parts of a coherent whole. Rather the diverse lived experiences of the building and the varied temporal, spatial, and social connections that were enacted through the different methods revealed a diverse and potentially contradictory multiplicity of realities [75]. Working with a combination of qualitative methods in a "methods assemblage" [68] allowed for this multiplicity, nuance, complexity, and dissonance to surface within the assessment process, and held those tensions without falling into incoherence or requiring their resolution through an artificial consensus. As Mol and Law observe, this messy complexity is often elided within official reports, with the objective tone that typifies much academic and professional writing, authoritatively establishing what is known, leaving limited space to reflect on the more unexpected and uncertain aspects of our knowledge [76] (p. 3).

Another advantage of adopting a mixture of qualitative methods is that they provide multiple avenues for participation. Differences in engagement across the methods proposed or adopted in this study highlights that methods are not equally accessible or appealing to participants. Such differences also emphasise the need for critical reflection on what individual methods might not reveal, or who might be unintentionally excluded from the process, as well as flagging practical considerations regarding where, when, and how people are willing and able to participate. As critics have argued, participatory processes do not inevitably result in greater inclusivity, empowerment, and sustainability; they can be co-opted or coercive, reinforcing existing practices and values rather than recognising the issues, knowledge, and spaces claimed by communities themselves [77,78]. Participatory processes designed to engage more marginalised groups are also open to 'capture' by the more advantaged and empowered middle class, who are familiar with the processes and terminology used in policy consultation [79]. Recognition of, and specific efforts to overcome, power differentials and existing inequalities, including critical self-reflection on personal positionality, values, and biases, are therefore essential to these processes.

"All action in the field of conservation is affected by an appraisal of value" [2] (p. 6). Perceptions of value determine what is done and how, as well as who decides and based on which forms of knowledge. In a Scottish context, the values that can be considered as part of a listing process are limited to assessments of architectural or historic interest, as determined by the underlying legislation. Nonetheless, public participation is an increasingly important part of heritage conservation, as professionals seek to balance the physical preservation of historic fabric with contemporary uses and values. In this context, the qualitative, participatory methods applied in assessing social values, and the nuanced understandings of the relationships between people and place that result, are arguably extremely useful and likely to become ever more relevant in the conservation and management of built heritage. However, bringing the knowledge resulting from a social values assessment into conservation practice is not without its challenges. Determining how pluralistic understandings of value can be incorporated into practical heritage management and conservation contexts remains "probably the most significant issue facing contemporary heritage management and policy" [26] (p. 1). As Macdonald observes, professionals are "only beginning to grapple with the implications for conservation in terms of which values take priority and how they are conserved" [80] (p. 7).

This case study has demonstrated that a mix of rapid, qualitative, participatory methods, deployed in a flexible, responsive, reflexive, and ethical manner, is practical within 'real world' conservation and heritage management contexts. However, to be most effective, participatory methods need to be embedded in genuinely people-centred heritage management and conservation processes. This means going beyond consultative approaches and engaging people in "invited spaces" [81] (p. 230) to recognise other forms of knowledge and expertise regarding what makes built heritage valuable. Jones and Yarrow [82,83] have described the collaborative processes and negotiations that take place between conservation practitioners, but community expertise and social values are rarely included in these processes. Working with participatory methods and engaging in truly participatory processes is an important step in opening up heritage and conservation decision-making, making otherwise hidden professional judgements and values more visible. Doing so also offers the potential for more inclusive and socially relevant forms of practice to emerge. The result is a more complex, but also a far richer, understanding of our historic environment and the contribution built heritage makes to people's lives.

Funding: This research was funded by the University of Stirling and Historic Environment Scotland through a MATCH collaborative PhD studentship.

Institutional Review Board Statement: The research was reviewed and approved by the General University Ethics Panel of the University of Stirling (Project identification code: GUEP434, dated 2 July 2018; subsequent amendment related to this case study, dated 5 April 2019).

Informed Consent Statement: All subjects gave their informed consent before they participated in the study.

Data Availability Statement: Due to participant confidentiality, the original data from this study has not been placed in a in a publicly accessible repository. Further details of the social values assessment are provided in the site report for Cables Wynd House, available via the University of Stirling Online Research Repository, https://storre.stir.ac.uk/ (accessed on 30 June 2023), and the discussion of the case study on the Social Value Toolkit, https://socialvalue.stir.ac.uk/ (accessed on 30 June 2023).

Acknowledgments: Full acknowledgement and grateful thanks are given to all the individuals who participated in this study; without their generosity in sharing their time and knowledge this research could not have taken place. I would also like to thank the designations team at Historic Environment Scotland, for their support in understanding the listing process for Cables Wynd House; my PhD supervisors, Siân Jones and Peter Matthews at the University of Stirling, and Judith Anderson and Karen Robertson at Historic Environment Scotland, for their guidance and support throughout this study and my wider doctoral journey; the colleagues who reviewed initial drafts of this paper; and the anonymous peer reviewers.

Conflicts of Interest: The author declares no conflict of interest.

References

1. Jones, S. Wrestling with the Social Value of Heritage: Problems, Dilemmas and Opportunities. *J. Community Archaeol. Herit.* **2017**, *4*, 21–37. [CrossRef]
2. Bell, D. *TAN [Technical Advisory Note] 08—The Historic Scotland Guide to International Conservation Charters*; Historic Scotland: Edinburgh, Scotland, 1997.
3. Jones, S. *Early Medieval Sculpture and the Production of Meaning, Value and Place: The Case of Hilton Cadboll*; Historic Scotland: Edinburgh, Scotland, 2004.
4. Jones, S.; Leech, S. *Valuing the Historic Environment: A Critical Review of Existing Approaches to Social Value*; Arts and Humanities Research Council: Manchester, UK, 2015.
5. Australia ICOMOS. *Guidelines for the Conservation of Places of Cultural Significance ("Burra Charter")*; Australia ICOMOS: Burra, Australia, 1979. Available online: https://australia.icomos.org/wp-content/uploads/Burra-Charter_1979.pdf (accessed on 30 June 2023).
6. Emerick, K. *Conserving and Managing Ancient Monuments: Heritage, Democracy, and Inclusion*; Boydell & Brewer: Woodbridge, UK, 2014.
7. Lesh, J. Social value and the conservation of urban heritage places in Australia. *Hist. Environ.* **2019**, *31*, 42–62.
8. Australia ICOMOS. *The Burra Charter: The Australia ICOMOS Charter for Places of Cultural Significance*; Australia ICOMOS: Burwood, Australia, 2013. Available online: https://australia.icomos.org/wp-content/uploads/The-Burra-Charter-2013-Adopted-31.10.2013.pdf (accessed on 30 June 2023).
9. Walker, M. The Development of the Australia ICOMOS Burra Charter. *APT Bull. J. Preserv. Technol.* **2014**, *45*, 9–16.
10. Council of Europe. *Council of Europe Framework Convention on the Value of Cultural Heritage for Society (CETS No. 199)*; Council of Europe: Faro, Portugal, 2005.
11. Drury, P.; McPherson, A. *Conservation Principles, Policies and Guidance for the Sustainable Management of the Historic Environment*; English Heritage: London, UK, 2008.
12. Historic Environment Scotland. *Historic Environment Policy for Scotland*; Historic Environment Scotland: Edinburgh, Scotland, 2019.

13. Belfiore, E.; Bennett, O. Rethinking the Social Impacts of the Arts. *Int. J. Cult. Policy* **2007**, *13*, 135–151. [CrossRef]
14. Byrne, D.; Brayshaw, H.; Ireland, T. *Social Significance: A Discussion Paper*; NSW National Parks and Wildlife Service: Hurstville, Australia, 2001.
15. Harrison, R. *Heritage: Critical Approaches*; Taylor and Francis: London, UK, 2013.
16. Lowenthal, D. *The Past is a Foreign Country*; Cambridge University Press: Cambridge, UK, 1985.
17. Hall, S. Whose Heritage? Un-settling 'The Heritage', re-imagining the post-nation. In *The Heritage Reader*; Fairclough, G., Harrison, R., Jameson, J., Jr., Schofield, J., Eds.; Routledge: Abingdon, UK, 2008; pp. 219–228.
18. Byrne, D. Heritage as Social Action. In *The Heritage Reader*; Fairclough, G., Harrison, R., Jameson, J., Jr., Schofield, J., Eds.; Routledge: Abingdon, UK, 2008; pp. 149–173.
19. Smith, L. *Uses of Heritage*; Routledge: Abingdon, UK, 2006.
20. Waterton, E. *Politics, Policy and the Discourses of Heritage in Britain*; Palgrave Macmillan: Basingstoke, UK, 2010.
21. Waterton, E.; Smith, L. There is No Such Thing as Heritage. In *Taking Archaeology Out of Heritage*; Waterton, E., Smith, L., Eds.; Cambridge Scholars Publishing: Newcastle upon Tyne, UK, 2009; pp. 10–27.
22. Carman, J. Where the Value Lies: The importance of Materiality to the Immaterial Aspects of Heritage. In *Taking Archaeology Out of Heritage*; Waterton, E., Smith, L., Eds.; Cambridge Scholars Publishing: Newcastle upon Tyne, UK, 2009; pp. 192–208.
23. Jones, S. Negotiating Authentic Objects and Authentic Selves: Beyond the Deconstruction of Authenticity. *J. Mater. Cult.* **2010**, *15*, 181–203. [CrossRef]
24. Avrami, E.; Mason, R.; de la Torre, M. *Values and Heritage Conservation: Research Report*; The Getty Conservation Institute: Los Angeles, CA, USA, 2000.
25. Clark, K. Values in cultural resource management. In *Heritage Values in Contemporary Society*; Smith, G., Messenger, P., Soderland, H., Eds.; Left Coast Press: Walnut Creek, CA, USA, 2010; pp. 89–99.
26. Gibson, L.; Pendlebury, J. Introduction: Valuing Historic Environments. In *Valuing Historic Environments*; Gibson, L., Pendlebury, J., Eds.; Routledge: Abingdon, UK, 2009; pp. 1–16.
27. Chitty, G. Introduction: Engaging Conservation—Practising Heritage Conservation in Communities. In *Heritage, Conservation and Communities: Engagement, Participation and Capacity Building*; Chitty, G., Ed.; Routledge: Abingdon, UK, 2017; pp. 1–14.
28. Emerick, K. The Language Changes but Practice Stays the Same: Does the Same Have to be True for Community Conservation? In *Heritage, Conservation and Communities: Engagement, Participation and Capacity Building*; Chitty, G., Ed.; Routledge: Abingdon, UK, 2017; pp. 65–77.
29. Pendlebury, J. *Conservation in the Age of Consensus*; Routledge: Abingdon, UK, 2009.
30. Walter, N. Everyone Loves a Good Story: Narrative, Tradition and Public Participation in Conservation. In *Heritage, Conservation and Communities: Engagement, Participation and Capacity Building*; Chitty, G., Ed.; Routledge: Abingdon, UK, 2017; pp. 50–64.
31. Waterton, E.; Smith, L.; Campbell, G. The Utility of Discourse Analysis to Heritage Studies: The Burra Charter and Social Inclusion. *Int. J. Herit. Stud.* **2006**, *12*, 339–355. [CrossRef]
32. Deacon, H.; Smeets, R. Authenticity, Value and Community Involvement in Heritage Management under the World Heritage and Intangible Heritage Conventions. *Herit. Soc.* **2013**, *6*, 129–143. [CrossRef]
33. Johnston, C. *What is Social Value?* Australian Government Publishing Service: Canberra, Australia, 1992.
34. Nara + 20: On Heritage Practices, Cultural Values, and the Concept of Authenticity. *Herit. Soc.* **2015**, *8*, 144–147. [CrossRef]
35. Sørensen, M.; Carman, J. (Eds.) *Heritage Studies: Methods and Approaches*; Routledge: Abingdon, UK, 2009.
36. Madgin, R.; Lesh, J. (Eds.) *People-Centred Methodologies for Heritage Conservation: Exploring Emotional Attachments to Historic Urban Places*, 1st ed.; Routledge: Abingdon, UK, 2021.
37. Taplin, D.H.; Scheld, S.; Low, S. Rapid Ethnographic Assessment in Urban Parks: A Case Study of Independence National Historical Park. *Hum. Organ.* **2002**, *61*, 80–93. [CrossRef]
38. Witcomb, A.; Buckley, K. Engaging with the future of 'critical heritage studies': Looking back in order to look forward. *Int. J. Herit. Stud.* **2013**, *19*, 562–578. [CrossRef]
39. Byrne, D. Counter-mapping in the Archaeological Landscape. In *Handbook of Landscape Archaeology*; David, B., Thomas, J., Eds.; Routledge: Abingdon, UK, 2008; pp. 609–616.
40. Harrison, R. 'Counter-Mapping' Heritage, Communities and Places in Australia and the UK. In *Local Heritage, Global Context: Cultural Perspectives on Sense of Place*; Schofield, J., Szymanski, R., Eds.; Ashgate Publishing Limited: Farnham, UK, 2011; pp. 79–98.
41. Jones, S.; Jeffrey, S.; Maxwell, M.; Hale, A.; Jones, C. 3D heritage visualisation and the negotiation of authenticity: The ACCORD project. *Int. J. Herit. Stud.* **2018**, *24*, 333–353. [CrossRef]
42. Schofield, T.; Foster Smith, D.; Bozoglu, G.; Whitehead, C. Design and Plural Heritages: Composing Critical Futures. In Proceedings of the 2019 CHI Conference on Human Factors in Computing Systems, Glasgow, Scotland, 4–9 May 2019.
43. Jeffrey, S.; Jones, S.; Maxwell, M.; Hale, A.; Jones, C. 3D visualisation, communities and the production of significance. *Int. J. Herit. Stud.* **2020**, *26*, 885–900. [CrossRef]
44. Robson, E. Wrestling with Social Value: An Examination of Methods and Approaches for Assessing Social Value in Heritage Management and Conservation. Ph.D. Thesis, University of Stirling, Stirling, Scotland, 2021. Available online: http://hdl.handle.net/1893/33355 (accessed on 30 June 2023).
45. Historic Environment Scotland. *Designation Policy and Selection Guidance*; Historic Environment Scotland: Edinburgh, Scotland, 2019.

46. Parliament of the United Kingdom. *Planning (Listed Buildings and Conservation Areas) (Scotland) Act 1997*; Parliament of the United Kingdom: London, UK, 1997. Available online: https://www.legislation.gov.uk/ukpga/1997/9 (accessed on 30 June 2023).
47. Scotland's Census: Area Profiles 2011. Available online: https://www.scotlandscensus.gov.uk/ods-web/area.html (accessed on 30 June 2023).
48. Scottish Index of Multiple Deprivation 2020. Available online: https://simd.scot/ (accessed on 30 June 2023).
49. Glendinning, M.; Muthesius, S. *Tower Block: Modern Public Housing in England, Scotland, Wales and Northern Ireland*; Yale University Press: New Haven, CT, USA, 1993.
50. 300008752_Designations Report on Handling_Decision. Available online: http://portal.historicenvironment.scot/decision/5000 01182 (accessed on 30 June 2023).
51. Ingold, T. *The Perception of the Environment: Essays on Livelihood, Dwelling and Skill*; Routledge: Abingdon, UK, 2011.
52. Davies, C.A. *Reflexive Ethnography: A Guide to Researching Selves and Others*, 2nd ed.; Routledge: London, UK, 2008.
53. Gillham, B. *Case Study Research Methods*; Continuum: London, UK, 2000.
54. Bryman, A. *Social Research Methods*, 5th ed.; Oxford University Press: Oxford, UK, 2016.
55. Pendlebury, J.; Townshend, T.; Gilroy, R. Social Housing as Heritage: The Case of Byker, Newcastle upon Tyne. In *Valuing Historic Environments*; Gibson, L., Pendlebury, J., Eds.; Routledge: Abingdon, UK, 2009; pp. 179–200.
56. Putnam, R. *Bowling Alone: The Collapse and Revival of American Community*; Simon & Schuster Paperbacks: New York, NY, USA, 2000.
57. Forrest, R.; Kearns, A. Social Cohesion, Social Capital and the Neighbourhood. *Urban Stud.* **2001**, *38*, 2125–2143. [CrossRef]
58. Douglas, M. *Purity and Danger*; Routledge: London, UK, 2002; First version 1966 published.
59. Cohen, A. A Sense of Time, A Sense of Place: The Meaning of Close Social Association in Whalsay, Shetland. In *Belonging: Identity and Social Organisation in British Rural Cultures*; Cohen, A., Ed.; Manchester University Press: Manchester, UK, 1982; pp. 21–49.
60. Macdonald, S. *Reimagining Culture: Histories, Identities and the Gaelic Renaissance*; Berg: Oxford, UK, 1997.
61. Malpass, P. Whose Housing Heritage? In *Valuing Historic Environments*; Gibson, L., Pendlebury, J., Eds.; Routledge: Abingdon, UK, 2009; pp. 201–214.
62. de Certeau, M. *The Practice of Everyday Life*; Rendall, S., Translator; University of California Press: Berkeley, CA, USA, 1984.
63. Yu, J.; Cooper, H. A Quantitative Review of Research Design Effects on Response Rates to Questionnaires. *J. Mark. Res.* **1983**, *20*, 36–44. [CrossRef]
64. Cooke, S.; Buckley, K. Visual research methodologies and the heritage of 'everyday' places. In *People-Centred Methodologies for Heritage Conservation: Exploring Emotional Attachments to Historic Urban Places*; Madgin, R., Lesh, J., Eds.; Routledge: Abingdon, UK, 2021; pp. 143–155.
65. Harper, D. Talking about pictures: A case for photo elicitation. *Vis. Stud.* **2002**, *17*, 13–26. [CrossRef]
66. Radley, A. What people do with pictures. *Vis. Stud.* **2010**, *25*, 268–279. [CrossRef]
67. Pink, S. *Doing Visual Ethnography: Images, Media, and Representation in Research*, 2nd ed.; Sage Publications: London, UK, 2007.
68. Law, J. *After Method: Mess in Social Science Research*; Routledge: Abingdon, UK, 2004.
69. Braun, V.; Clarke, V. Using thematic analysis in psychology. *Qual. Res. Psychol.* **2006**, *3*, 77–101. [CrossRef]
70. Beebe, J. Basic Concepts and Techniques of Rapid Appraisal. *Hum. Organ.* **1995**, *54*, 42–51. [CrossRef]
71. Pink, S.; Morgan, J. Short-Term Ethnography: Intense Routes to Knowing. *Symb. Interact.* **2013**, *36*, 351–361. [CrossRef]
72. Pink, S. *Doing Sensory Ethnography*; SAGE Publications Limited: London, UK, 2009.
73. Silverman, M. Everyday Ethics. A Personal Journey in Rural Ireland, 1980–2001. In *The Ethics of Anthropology: Debates and Dilemmas*; Caplan, P., Ed.; Routledge: London, UK, 2003; pp. 115–132.
74. Douglas-Jones, R.; Hughes, J.; Jones, S.; Yarrow, T. Science, value and material decay in the conservation of historic environments. *J. Cult. Herit.* **2016**, *21*, 823–833. [CrossRef]
75. Mol, A. *The Body Multiple: Ontology in Medical Practice*; Duke University Press: Durham, NC, USA, 2002.
76. Mol, A.; Law, J. Complexities: An Introduction. In *Complexities: Social Studies of Knowledge Practices*; Law, J., Mol, A., Eds.; Duke University Press: Durham, NC, USA, 2002; pp. 1–22.
77. Cruikshank, B. *The Will to Empower: Democratic Citizens and Other Subjects*; Cornell University Press: Ithaca, NY, USA, 1999.
78. Cooke, B.; Kothari, U. The Case for Participation as Tyranny? In *Participation: The New Tyranny?* Cooke, B., Kothari, U., Eds.; Zed Books: London, UK, 2001; pp. 1–15.
79. Matthews, P.; Hastings, A. Middle-Class Political Activism and Middle-Class Advantage in Relation to Public Services: A Realist Synthesis of the Evidence Base. *Soc. Policy Adm.* **2013**, *47*, 72–92. [CrossRef]
80. Macdonald, A. Modern Heritage in the Twenty-first Century. *Conserv. Perspect. GCI Newsl.* **2023**, *38*, 4–9.
81. Taylor, M. *Public Policy in the Community*; Palgrave Macmillan: Basingstoke, UK, 2011.
82. Jones, S.; Yarrow, T. Crafting Authenticity: An Ethnography of Conservation Practice. *J. Mater. Cult.* **2013**, *18*, 3–26. [CrossRef]
83. Jones, S.; Yarrow, T. *The Object of Conservation*; Routledge: London, UK, 2022.

Disclaimer/Publisher's Note: The statements, opinions and data contained in all publications are solely those of the individual author(s) and contributor(s) and not of MDPI and/or the editor(s). MDPI and/or the editor(s) disclaim responsibility for any injury to people or property resulting from any ideas, methods, instructions or products referred to in the content.

Article

Towards a Holistic Narration of Place: Conserving Natural and Built Heritage at the Humble Administrator's Garden, China

Youcao Ren [1] and Johnathan Djabarouti [2,*]

1   Department of Architecture, Sheffield School of Architecture, Sheffield S10 2TN, UK; youcao.ren@sheffield.ac.uk
2   Department of Architecture, Manchester School of Architecture, Manchester M1 7ED, UK
*   Correspondence: j.djabarouti@mmu.ac.uk

Abstract: World Heritage tourism in China regulates conservation approaches employed across natural and built heritage sites. However, focusing on the revenue-generating potential of these sites sustains material authenticity and technical conservation methods. The outcome is a conflict between conservation and commercialization, where socio-cultural values are overshadowed by the process of museumization. Underpinned by critical heritage theory and a focus on intangible heritage, this research seeks to confront this conflict by examining the shifting conservation practice at the Humble Administrator's Garden (HAG), a World Heritage Site and Classical Garden of Suzhou, China. A mixed-methodological approach explores the interplay between architecture and landscape within its heritage conservation process, utilizing archival research, semi-structured interviews with HAG Management, and visitor journals. The study shows how HAG's heritage is shaped by visitors' personal experiences and emotions alongside expert interpretations, resulting in the foregrounding of diverse narratives that contribute to a holistic sense of place. Within its politicized system, the safeguarding of intangible heritage requires constant negotiation among the municipality, the market, and emerging narrators. Attempts to reinterpret its former heritage buildings demonstrate a changing conservation discourse as the site transitions from an exclusive literati estate to a multivocal space of cultural encounter. The study illustrates how a focus on narrative representation unifies architecture and landscape, reimagining centuries of literati culture. This makes conceptual space for considering how conservation management can inform a more holistic narration of 'place' at similar World Heritage sites via the foregrounding of previously silent stakeholders.

Keywords: cultural landscapes; built heritage; conservation; intangible heritage; critical heritage theory

Citation: Ren, Y.; Djabarouti, J. Towards a Holistic Narration of Place: Conserving Natural and Built Heritage at the Humble Administrator's Garden, China. *Architecture* **2023**, *3*, 446–460. https://doi.org/10.3390/architecture3030024

Academic Editor: Avi Friedman

Received: 23 June 2023
Revised: 2 August 2023
Accepted: 9 August 2023
Published: 14 August 2023

Copyright: © 2023 by the authors. Licensee MDPI, Basel, Switzerland. This article is an open access article distributed under the terms and conditions of the Creative Commons Attribution (CC BY) license (https://creativecommons.org/licenses/by/4.0/).

1. Introduction

World Heritage Site (hereafter WHS) inscription has been considered a catalyst for cultural heritage tourism development worldwide [1–3]. This is especially relevant in a Chinese context where the effect of WHS tourism can be clearly observed [4,5]. In transforming WHSs into tourism resources, the values and meanings of China's WHSs are often defined by and safeguarded through the contested input of the government, heritage agencies, and the general public [6]. However, tension between stakeholders can trigger displaced interpretations of heritage assets that require conservation methodologies to be constantly reappraised. Firstly, official Chinese mechanisms have attached political significance to WHS inscription by imposing pressure on heritage sites to maintain their material authenticity. Interpretative decision-making is commonly translated into strict monitoring systems and/or static heritage exhibitions. Secondly, WHSs have been widely utilized as catalysts for revenue over other social and cultural values [4]. Such a tendency primarily leads to the proliferation of commercially constructed tourism programs, which triggers heritage representation methodologies.

The Classic Gardens of Suzhou falls within the intersection of these two associated yet somewhat contested discourses. Once certified as 'authentic', classic gardens are decontextualized and displayed as relics for inspection and appreciation by the visiting public [7]. Consequently, the socio-cultural values of the gardens that support their designation as heritage tends to become obscured in the process of museumization. Efforts to catalyze the transformation of demarcated heritages into tourism resources, therefore, face challenging issues between traditional heritage conservation conflicting with the management of change to facilitate development processes [8].

Drawing on the anxieties imposed by critical heritage theory (in terms of heritage conceptualization and representation), we explore the shifting conservation practice undertaken at one of the first Classical Gardens to be inscribed as a WHS garden in 1997—the Humble Administrator's Garden (hereafter HAG) and its associated architecture (The Li Residence). This 16th century scholarly estate of the HAG is considered representative of the Classical Gardens, often portrayed as a physical manifestation of 'scholar-official' dominance across society, aesthetics, and building skills in imperial China [9]. Yet within this article, we assert that this portrayal is also underpinned by an interpretation of the HAG as a symbol of unity between architecture and landscape [10]. In reference to this, we explore how its conservation increasingly works towards nurturing a holistic narration of place, capturing both building and landscape within its conservation assessment and management. We further highlight how this fosters the communication of non-physical heritage (or 'intangible heritage') in the evolving conservation approach of the HAG.

2. Intangible Heritage and the Cultural Landscape

Since the middle of the 20th century, international heritage protection mechanisms developed by UNESCO and ICOMOS have been critiqued for their restricted conceptualization of heritage [11] (p. 27), [12] (p. 18), with their focus being placed primarily on physical art-historical manifestations of national identity [13] (p. 18). The creation of the World Heritage Convention (hereafter WHC) [14] was one of the first clear international moves towards a broader definition of cultural heritage, with the landscape being formally established as a definitive cultural heritage category [15] (p. 39). This resulted in additional focus given to areas and places that contained 'natural features', 'geological and physiographical formations', and 'natural sites' [14] (p. 2). Subsequently, this paved the way for more nuanced terminology and guidance in relation to landscape, such as Historic Gardens [16], and eventually the term 'cultural landscape', which became formalized in the WHC in an early 1990s revision [17] (p. 122). This expansion of cultural heritage can also be seen to track a broadening of alternative conservation ideologies [18] (p. 7). Although the concept of cultural landscapes was conclusively integrated into world heritage terminology over thirty years ago, the general idea of natural heritage is still comparatively new to the concept of World Heritage [19] (p. 159). This is despite the broader ideas and sentiments of natural landscape protection being ingrained within both the Romantic Movement and overarching Enlightenment philosophy [13] (p. 21), [18] (p. 3). Interestingly these are two key historical moments that assisted in defining the principles of what would become the modern conservation movement in the late 19th century.

Whilst heritage categories expanded, a sentiment of disapproval over a lack of representation within UNESCO's worldview of heritage continued and reached a crescendo in the 1990s [20] (p. 97), when UNESCO itself began to acknowledge the inherent precedence given to West European material heritage sites at the expense of 'living' cultural heritage. This awareness was buttressed by criticism from heritage scholars and non-Western UNESCO member states who lobbied for a more inclusive concept of World Heritage [21] (p. 32). The Convention for the Safeguarding of the Intangible Cultural Heritage [22]—commonly referred to as the Intangible Cultural Heritage Convention (hereafter ICHC)—is often regarded as an outcome of this lobbying, although the heritage value of cultural landscapes was given greater focus in the decade leading up to the production of the ICHC [13] (p. 31).

Of particular interest to this contribution is the connection and synergy between heritage buildings, cultural landscapes, and intangible cultural heritage (hereafter ICH), with the use and evolution of cultural landscapes being actively transformed through the practices that are sustained by intangible heritage [15] (p. 40). The ICHC hoped to offer a remedy for the unfair distribution of World Heritage [23] (p. 964) by initiating an official definition of ICH (hereafter ICH) [24] (p. 919). Formalized guidance for its safeguarding was also introduced, which avoided nomenclature typically used to describe the significance of tangible heritage [25] (p. 75). UNESCO defines ICH within Article 2 of the 2003 Convention as

> *...the practices, representations, expressions, knowledge, skills... that communities, groups and, in some cases, individuals recognize as part of their cultural heritage. This intangible cultural heritage... is constantly recreated by communities and groups in response to their environment, their interaction with nature and their history, and provides them with a sense of identity and continuity...* [22] (p. 2)

Lenzerini [26] (p. 101) summarizes the key themes of the ICHC as self-identification; constant re-creation; identity; authenticity; and human rights. Thus, the convention is known for being constructed to address globalization through the support and celebration of cultural diversity [22] (p. 1), [27] (p. 265). Ironically, it is this creeping scope of what constitutes World Heritage that has also led to increasing control by states/governments over a much broader range of heritage typologies [17] (p. 56). Indeed, from a tourist perspective, the convention could be utilized as a tool to misuse those very practices that it has been created to guard by raising local traditions onto global platforms [28] (p. 78), [29] (p. 731), [30]. The potentially confusing result is that UNESCO becomes the creator of the very issue they are seeking to resolve through the production of the ICHC [17] (p. 115), [27] (p. 266).

3. A Holistic Understanding of 'Place'

Increasing interest in an integrated understanding of World Heritage and intangible qualities of heritage is having an impact on how much cultural landscapes are acknowledged when considering the conservation of built heritage. There are two key reasons for this. Firstly, as ICH primarily concentrates on practices, the implication of this focus is an interest in the various types of landscapes where these practices take place [17] (p. 115). Thus its impact has been to confront the "nature–culture split" that Hill [31] describes as central to the construction of traditionally siloed heritage domains. Through the merger of natural and cultural concepts, the landscape is reconceptualized as heritage and subsequently entangled in the idea of 'place' alongside buildings, monuments, and other artifacts [16] (p. 31), [32]. Secondly, as the notion of heritage expands to incorporate new models of what heritage could be (landscape and historic gardens being examples of one such expansion route), any built heritage site appraisal/assessment must now take appropriate notice of the broader site and landscape within which it is contextualized (i.e., within a Heritage Impact Assessment at a World Heritage Site). This approach is most revealing within the Burra Charter's practitioner guidance—in particular, their definition of places of cultural significance as being comprised of "...elements, objects, spaces and views... [and] may have tangible and intangible dimensions" [32]. Therefore, when considering the physical conservation of built heritage fabric, professional decision-making will be influenced by a concern for the cultural heritage landscape in a more holistic and balanced sense than has previously been the case. For example, in a study concerning the Class II designated Hilton of Cadboll stone in Scotland, UK, Jones [20] (p. 105) describes how a more integrative reading of the relationship between landscape, historical monuments, and people can holistically constitute 'place' and therefore blur the differentiation between building and landscape. However, where this remains problematic for many World Heritage Sites is how the landscape is often conceived as a backdrop or complementary addition to architectural and/or urban tangible heritage (for example, see [33] (p. 63)).

In relation to this, it is worth re-visiting the now timeless statement by Laurajane Smith, which, whilst many use this to form critical positions with regards to the conservation of built heritage, can also be seen to have relevance to cultural heritage landscapes as well:

> *It is value and meaning that is the real subject of heritage preservation and management processes, and as such all heritage is 'intangible' whether these values or meanings are symbolized by a physical site, place, landscape or other physical representation...* [13] (p. 56)

When conceptually processed through the notion of 'heritage' as a reflexive present-day process [34], landscapes become both cultural and mnemonic [13] (p. 46), [35] (p. 5). Pierre Nora's term *milieux de memoir* (environments of memory) reflects this, whereby a landscape becomes part of ongoing present-day interactions to create an active "memory culture" see [36] (p. 27). Hence the emergence of terms that seek to add a cultural deposit to the term landscape—such as "memoryscape" [37], "memorialscape" [38], "deathscape" [36] (p. 36), and so on. In considering the heritage value and meaning of a cultural heritage landscape, Smith [13] (p. 56) asserts that conceptualizing heritage as "intangible" shifts concerns to that of heritage effect, which is the way spatial configurations and practices can move people in various ways (some examples given include emotionally, politically, and culturally). So even more than built heritage, cultural heritage landscapes can accommodate various contemporary interpretations of history and society, with many of these understandings often being in direct conflict with one another [39] (p. 89), [40] (p. 6). A cultural landscape can therefore be understood as culturally, experientially, and emotionally layered [41] (p. 42), inclusive of the tangible heritage that is situated upon it.

4. Evolving from a Traditional to a Contemporary Conservation Approach

At the core of built heritage conservation is a quest for the representation of historical truth in physical remains that are deemed to be of value [42] (p. 28). A traditional conservation approach typically achieves this through the communication of a prevailing narrative to the users of the heritage, and the concern for centuries has been how conservative or innovative one can be when considering how to communicate and/or represent this narrative. Conversely, a contemporary conservation approach—which is situated within the context of a postmodern heritage paradigm—must now evidence diversity and multi-vocality by demonstrating a steady evolution of significances (note the plural) that track contemporary values [11] (p. 15), [42] (p. 2). Whilst a cultural heritage landscape has the capacity to accommodate this notion of re-evaluation, built heritage is far more problematic—primarily because it is both philosophical and logistically rigid. Equally, as already touched upon, cultural heritage places must also now be malleable enough to represent and uphold a variety of oftentimes conflicting narratives, meaning different conservation methods may be required to represent different stories, and therefore heritage practitioners will likely require shifts in appraisal and management approaches to better reflect heritage as non-physical, processual, and dynamic [43] (pp. 1110–1111). These shifts that are required will not only be related to the development of approaches that push beyond the classic binary of preservation and restoration but also to how traditional approaches are valued and viewed in contemporary society (for example, see Djabarouti [44] (p. 122)). These differing approaches have been described as the 'tautological argument' of restoration [18] (p. 208), where the history of the building must be rationalized against its present-day development, which paradoxically also becomes history itself via the passage of time. The question for built heritage practitioners is, therefore, whether one should utilize a conservation method that can venerate a particular moment in time or should a multiplicity of times and stages of development be conserved [45] (p. 4).

When considering world heritage in the more holistic sense (landscape, built form, and the cultural practices that shape them), this question becomes even more obscured, which serves to demonstrate how contemporary understandings of heritage are moving faster than the conservation appraisal and management approaches that practitioners employ. Hence why international heritage sites that capture both landscape (dynamic) and building (static), as well as representations of both tangible (physical) and intangible (non-physical) heritage, such as the HAG, serve as useful cases for exploring this interplay between building and landscape in more detail and through a heritage- and place-specific lens. Furthermore, as the heritage scope broadens to consider this interplay, conservation concerns of a more performative nature emerge in terms of how a place can support society to reproduce commemorative ceremonies, bodily practices, and other everyday practices through social performance [46] (p. 23), [47] (p. 10).

5. Materials and Methods

The scope of our case study exploration primarily concerns two aspects. Firstly, addressing the physical and cultural divisions of architecture and landscape within the HAG influences its integrity as a WHS and the sustainability of its changing socio-cultural significance. Secondly, exploring the experimental narrative development process that now foregrounds previously silent heritage stakeholders. We assert that the transitioning interplay between the government, conservation practitioners, and the public in representing this WHS is reflective of the changing perception of heritage in a Chinese context, which is inherently linked to a globalized critical re-evaluation of how heritage is conceptualized.

The case of the HAG is explored via a mixed methodological approach. Case study data was obtained from archival research. Policy papers, reports, and conservation records of the HAG since the 1940s were accessed through public archives and the HAG Management Office. We then conducted semi-structured interviews with the HAG Management Office between February and April 2023. Each interview lasted between 40–90 min to allow in-depth discussions that revolved around three themes: the HAG's responses to WHS conservation policy; the interviewee's role in intervention; and the identification of involved actors in the HAG representation process. In addition, we also reviewed 52 visitor journals that were collected between March and May 2023. This offers a set of triangulated data sources, the benefit of which being to not only verify the data [48] but, more specifically for this case, to ensure that differing (and possibly opposing) views are consolidated into the study findings in relation to the interplay between the buildings and landscapes on the site [49] (p. 81). Visitor journals accessed through the HAG Management Office were anonymized upon collection.

6. Case Study: (Re)Conceptualizing the Humble Administrator's Garden
6.1. Overview

The culturally and socially significant HAG is considered a representative of the Classic Garden of Suzhou and was among the first to be inscribed as a WHS garden in 1997. Completed at the peak of imperial China, the estate is conceptualized by both a prevailing scholar-official culture in China's East Yangtze Delta region and an aesthetic of traditional Chinese landscape painting. It conveyed the most esteemed pursuits of contemporary Chinese society, such as the joie de vivre of home (architecture) being closely integrated with nature (landscape) and social status being achieved through the attainment of a political career (Figure 1). The housing complexes situated at the southern end of the site are representative of typical Suzhou residential architecture, whilst, in the north, the significance given to the garden results in a water-based naturalistic landscape that is embellished by buildings. In this study, the HAG is considered a symbol of unity between architecture and landscape [10]. Within this context, this contribution focuses on exploring the blurred boundary between architecture and landscape to illustrate how the HAG fosters a holistic narration of place, capturing both building and landscape within its conservation processes.

Architecture, therefore, serves as the key medium through which place-making is not only about the physical manifestation of space but also about the ideology and culture that underpins it. UNESCO's synthesis of the Classic Garden of Suzhou's Outstanding Universal Value (hereafter OUV) reminds us acutely of the significance of this holistic approach. This has implications that extend beyond academic discussion to encompass long-term conservation strategies that are inclined heavily toward landscape:

> ... *These garden ensembles of buildings, rock formations, calligraphy, furniture, and decorative artistic pieces serve as showcases of the paramount artistic achievements of the East Yangtze Delta region; they are in essence the embodiment of the connotations of traditional Chinese culture.*
>
> *Criterion (iv): The classical gardens of Suzhou are the most vivid specimens of the culture expressed in landscape garden design from the East Yangtze Delta region in the 11th to 19th centuries. The underlying philosophy, literature, art, and craftsmanship shown in the architecture, gardening as well as the handcrafts reflect the monumental achievements of the social, cultural, scientific, and technological developments of this period.*
>
> *Criterion (v): These classical Suzhou gardens are outstanding examples of the harmonious relationship achieved between traditional Chinese residences and artfully contrived nature. They showcase the lifestyle, etiquette and customs of the East Yangtze Delta region during the 11th to 19th centuries.*

Our investigation of the HAG, therefore, begins first with understanding the impact of a transformational conservation approach in supporting the notion of unifying architecture and landscape. To explore this in more detail, a more specific focus will be placed on the HAG's architecture complex, namely, the Li Residence, and its transition from a static exhibit to a space of cultural encounter. The Li Residence is the HAG's sole remaining housing complex that consists of a series of typical East Yangtze Delta timber-framed, multi-storey dwellings, with its plan distributed along two north–south parallel axes.

Figure 1. The Humble Administrator's Garden as drawn by Wen Zhengming (act. ca. 1500–1535) when first developed in the 16th century. Note that architecture and everyday living were portrayed as inseparable from the landscape (Source: The Met Museum, https://www.metmuseum.org/art/collection/search/39654, accessed on 30 April 2023). Image use rights: Public Domain.

6.2. The Divided Garden

Over the past five centuries, the HAG has undergone multiple changes of ownership that have led to significant alterations to its physical fabric. The current appearance of the HAG dates from the late Qing Dynasty (1644–1911), during which time the housing complexes towards the south of the garden were divided into three parts. These were known as, in order from east to west: the Li Residence; the Residence of Prince Loyal; and the Zhang Residence. Each of these three complexes covers an area of approximately 2400 square meters, 7700 square meters, and 8000 square meters, respectively (Figure 2).

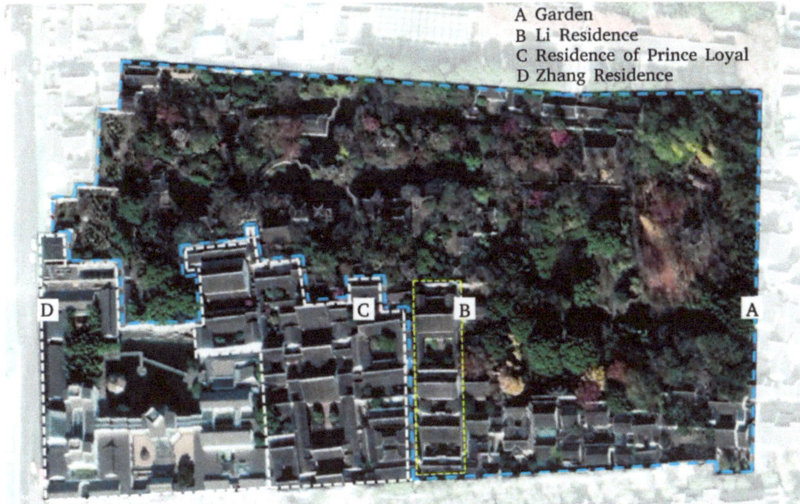

Figure 2. The Humble Administrator's Garden as of 2022. This unity of architecture and landscape involves the garden in the north and architecture complexes in the south. (Source: Google Earth, re-drawn by author).

Today it is almost impossible to trace with certainty whether the HAG's form in the late Qing was highly consistent with that of the 16th century due to the lack of historical plans. However, significant alterations seem likely based on the surviving records that indicate a continuous renewal of the architecture and landscape over the centuries. Renovations between the 19th and 20th centuries focused on the addition of ornamental garden buildings and the expansion of housing complexes, including the complex to the southwest of the garden that was later known as the Zhang Residence. The series of alterations that were made to the HAG, transforming both its housing complexes and the garden, were perceived as exclusively sympathetic to the owners' cultural interests, leading to public scrutiny. Thus, this process—rather than detracting from the HAG's socio-cultural significance—resultingly contributed to the public appreciation of scholar-official place-making activities. Through this, the HAG has continued to convey the aesthetic and lifestyle interests of the scholar-officials that influenced the architectural culture of the entire East Yangtze Delta region. Under this premise, the changing HAG remained authentic, although in a way that differs from the authenticity defined in Western conservation ideologies. The HAG's residential complexes were repurposed for various functions in the early communist era (1940s–1950s), which arguably led to a dramatic transition of its historical socio-cultural meanings. Whilst the Residence of Prince Loyal became an independent historical display building from the 1960s onwards, the Zhang Residence was repurposed for multiple municipal functions in succession, leading to its demolition in 2003 after decades-long deterioration. The site was later requisitioned for the development of Suzhou Museum, designed by architect I.M. Pei and opened in 2007. The Li Residence, on the other hand, was reformed to accommodate a local art school until the late 1980s. Corresponding to this is the reform of the historic private gardens prompted by strong contemporary socialist thinking, resulting in the HAG being opened to the public as an urban park in the 1950s. This conveyed a clear message from the local authority: the once scholarly landscape now belongs to the masses. The

HAG's previously integrated housing complexes and garden, in addition to being spatially divided, were given different functions respectively: the architecture was now to symbolize the mundane life of ordinary people, while the landscape was to serve as a contemporary public urban resource. The notion of a unity existing between architecture and landscape that once existed in the overarching scholar-official dominated narrative consequently ceased to exist.

7. Results

7.1. Restoring a Historical Unity

Whilst the primary scope of this paper revolves around an exploration of the HAG's historical evolution, it is also our intention to demonstrate how this process has led to a very predictable approach toward conservation methods and practice. Records from the HAG Management Office suggest approximately 1.2 million RMB of investment in the HAG's physical restoration at the end of the 20th century, which was concerned with ensuring the spatial form of both architecture and landscape was faithfully restored. Of the three complexes that once existed in the south of the HAG, only the Li Residence remains intact to this day. The WHS inscribed HAG as a unified housing complex and garden that covers an area of 52,000 square meters. The conservation of its housing complex (i.e., the Li Residence), however, contrasts markedly with that of the landscape. Between 1992 and 2007, the Li Residence was opened to the public as the first garden museum in China. Static displays were curated with the aim of communicating the values of the HAG through associative objects, with the architecture acting as the backdrop. The built heritage, in a rather literal sense, was museumized.

The caution in approaching material authenticity as the primary conservation focus is increasingly evident in the Li Residence's systematic restoration between 2009 and 2014, arguably to further align with UNESCO's framework of authenticity. Traditional conservation and construction techniques were applied in addition to reusing a considerable amount of original building material (anastylosis), which was supported by the involvement of traditional local craftspeople [50]. Yet despite its architectural restoration, the HAG continues to prioritize the garden as its main tourist resource. Upon its reopening in 2014, the now materially authentic Li Residence was repurposed from a museum gallery to the primary exit from the garden. It has since served as a residential showcase space for the HAG. This is primarily due to the Li Residence's limited capacity, with peak tourism traffic in the HAG reaching up to 9000 people per hour. The practice of utilizing a heritage building as a garden exit has legitimized an approach towards the Classical Gardens of Suzhou, with similar considerations being shared among other Classical Gardens. The built heritage that is intimately entangled with the history of the garden has been transformed from a space of encounter to a space of transit, where only unmeaningful experiences are likely to be accommodated at the end of the tours that take place in the gardens. The HAG's Management Office attempted to revive this fading sense of unity between architecture and landscape by embedding an immersive multimedia tour that was developed in 2020. This night tour program forms part of the Suzhou government's municipal strategy to restore the local economy in a post-COVID time. Here we have attempted to summarize the immersive multimedia tour—if not over-simplify it—as our focus is not the tour per se but the collaborative yet contested heritage commercialization process that it engenders.

Upon entering the Li Residence on the south end of the HAG, one's spatial movement follows a narrated route. The notion of walking is guided and framed along the central axis northwards until entering the garden. Throughout this experience, the route restores the HAG's historical spatial form of a house at the front and a garden at the back. The tour then continues and unfolds around the water body (Figure 3).

Reimagined intangible heritage—in this case, literati aesthetics, activities, and feelings that tie together the HAG's buildings and landscape, is translated into scenes. A total of eight scenes are designed within the Li Residence and fifteen in the garden, respectively. This description is perhaps somewhat paradoxical as the sense of rigid division between architecture and landscape is diluted in the tour. This is represented by, for instance, the imitations of scholarly activities that are centered around the notion of viewing. Building components—including walls, beams, and decorative elements—are used as part of the display to facilitate immersive multimedia experiences that depict garden life scenes (Figure 4).

Figure 3. The notion of walking is guided and framed along the Li Residence's central axis northwards until entering the garden. Blue line and arrows indicate the route. (Source: author re-drawing of an image provided by the HAG Management Office).

Figure 4. 'Sitting with the Cherry Trees'. Designed scenes in the Li Residence represent historical garden living and portray architecture as the agency of viewing. (Source: Image provided by the HAG Management Office). Permission of use granted.

The Li Residence's otherwise obscured function that extends beyond living is hereby unfolded. Through this, architecture is (re)portrayed as the agency between people and the environment. As one visitor's journal remarked:

I have visited (the HAG) before but this is a different experience. I suppose there is an aesthetic conception that's communicated through the scenes and the narrated tour. (VJ 6)

7.2. The Heritage Narrators

It is important first to acknowledge that the primary purpose of the HAG Management Office's participation in developing the immersive multimedia tour is to reinforce state control of heritage resources whilst maximizing local economic benefits [6]. The management of the Classical Gardens in Suzhou has fallen within the jurisdiction of the Suzhou Garden Management Office since its establishment in the early communist era. Affairs, including conservation and heritage commodification, were henceforth considered governmental matters. The role of conservation practitioners in negotiating the contested heritage value, tourism profits, and local policies remains relatively vague. It is also worth noting that the HAG is by no means the first heritage asset to embark on an attempt at performance tourism. Local WHS gardens have successful histories of manufacturing and simulating historical East Yangtze Delta region culture and scholar-official lifestyle to enhance tourist attraction. To give an example, the Master of Nets Garden, which was inscribed on the World Heritage List (hereafter WHL) in 1997, continues to accommodate night-time Kun Qu Opera performances in its garden. The major theatrical form of Kun Qu Opera was locally rooted (14th to 17th century) and was inscribed as ICH in 2021. Although there is no evident connection between the design and construction of the Master of Nets Garden and Kun Qu Opera, the program has been well received for (re)creating a garden life scene. Its long-lasting success has created a paradigm for Suzhou's WHS gardens. Both the Lion Forest Garden and the Canglang Pavilion, inscribed on the WHL in 1997 and 2000, respectively, have also applied a similar strategy. The immersive Kun Qu Opera performance in the Canglang Pavilion, for instance, is adapted from the life events of garden owners in the 19th century. The priority of the visitor experience is a purposefully designed narrative experience rather than simply the visual appreciation of physical heritage fabric (Figure 5).

Whilst performance tourism in a series of WHS gardens has proven successful in generating economic profit, the HAG Management Office had only established the immersive multimedia tour in 2020 as a response to the local authority's post-COVID economy recovery plan. A local tourism entrepreneur was entrusted to operate the HAG's immersive multimedia tour upon its creation in 2020, of which a 20 million RMB (approx. 3 million USD) investment was made. In 2021 HAG's immersive multimedia tour generated annual revenue of 6 million RMB. The entrepreneur, however, does not bear the responsibility of heritage safeguarding as a profit-driven third party. As a subordinate unit of the local garden bureau, the HAG Management Office led the project on the ground. The team's leading position is reflected in their power to certify what is appropriate to be included as part of the heritage interpretation process and how specific experiences should be translated into the immersive multimedia scene. Our interview with the HAG Management Office illustrates the overlapping roles of local government, conservation practitioners, and the HAG visitors in wrestling with the heritage commodification process, with one interviewee noting:

While the design of each scene is operated by our partner entrepreneur, our experience in the HAG every day feeds into the overall development of the tour. To us, the presentation of this tour is personalized. This distinguishes us from other WHS Suzhou gardens and WHSs across the state. (I 03)

Figure 5. The use of the WHS class garden as a venue for Kun Qu Opera performance is considered common practice in creating tourist attractions. (Source: author original image).

The HAG Management Office addresses its attempted narrative approach as an experimental representation that foregrounds physical and non-physical heritage, in contrast to the common practice among WHS gardens. It is, however, worth noting that both approaches are successful in generating attraction even though the programs are commodity-led initiatives. Whilst feelings of individual agency are embodied through their ongoing interaction with the external world [6,51,52], visitors interpret, consume, and negotiate the concepts and values of the Classic Garden of Suzhou with limited constraints imposed on them from the format of performative tourism. Visitors' journals of the HAG immersive multimedia tour illustrate vividly how their interpretations of heritage might echo (see VJ 11) or differ from (see VJ 3) the intended communication. This would also involve interpretations that, although as limited as 1 out of 52 reviewed journals, contest the intended outcome (VJ 39). We are particularly mindful of the fact that the representation of the previously overlooked architecture triggers a variety of feelings and understandings, where mundane activities are increasingly part of the narrative. The interpretation of HAG'S authenticity, therefore, is negotiated among these emerging stakeholders:

> *You can tell the differences between the HAG at night and during the day. I love the rain scene that is (re)created in the building. The rain, the banana trees, and the sound of thunder, together create a feeling of being there. (VJ 11)*

> *I am particularly intrigued by the architecture. Scenes in the Li Residence have created a peculiar Chinese horror atmosphere. I hope that there will be a live action role-playing game in the Li Residence. (VJ 3)*

> *To us, the tour is an imagination of the HAG's heritage. It does not feel authentic. (VJ 39)*

8. Discussion

As China's heritage commodification remains governed entirely by the local authorities [53], they, in turn, take administrative triumphs for successful outcomes. It is clear from the HAG case that intervention from heritage agencies in the heritage commodification process remains extremely limited. Rather, programs are entirely operated by the profit-driven private section, of which their participation in actual heritage safeguarding remains unclear.

The country's tourism-led conservation framework has therefore approached WHSs with a clear purpose: to serve the municipality. The primary starting point and purpose of heritage representation, therefore, remains heavily revolved around administrative and financial rewards. Within this context, heritage representation is considered an effective approach to generating tourism attraction and has ongoing effects on not only the HAG alone but a few of China's WHSs. Instances involve light display in the West Lake Cultural Landscape (Hangzhou), opera performance in the Forbidden City (Beijing), and folklore performance of Mogao Caves (Dunhuang), to name a few. Having achieved wide recognition, these representation attempts are incredibly resource hungry and further portray WHSs as local assets that compete with one other.

Where does this leave the narrators, then, when heritage representation is driven and operated by the municipality? The only way in which human beings can conceptualize self-identity is through narrative [54]. The complexity of creating 'narratives' in a landscape is critical in working towards a definition of heritage as an entanglement of dependencies between feelings and things [55,56]. Yet it is precisely because local authorities and the market have long aligned their interests within the heritage management system that limited space is left for credible professional interventions. For the HAG team, the tour as a heritage presentation outcome is set apart from other WHSs as their role extends to the narrator while leading the process. As multimedia scenes in the Li Residence resonate with the team's understanding of the building fabric and associated daily memories and feelings, one would be walking through not simply the building fabric but also their interpretations of the HAG. This demonstrates a subtle negotiation within the seemingly unbreakable government-market heritage management system. The presence of narrators, regardless of how implicit, is partially accountable for the manifestations of immaterial culture in practice at HAG as part of its shifting conservation practice.

The perception of the HAG's heritage does not necessarily require a 'proper' interpretation from the experts [13], despite the transitioning interplay between the government, conservation practitioners, and the public at the site. Rather, it is the entangled personal experiences, expressions, and emotions that produce stories that unify the material and immaterial heritage into a more holistic sense of place [57,58]. The visitor journals illustrate that the (re)designed HAG story triggers complex personal interpretations that expand the HAG's curated scenes into diverse narratives, although they seem to have been put in a position of obscurity in the current discourse. The depiction of everyday activities and feelings in the visitor journals reflects the HAG's shift from a site previously portrayed as transcendent to a site that supports society to reproduce commonplace practices through social performance. This reinforces the notion that conservation is fundamentally a process concerned with preserving and enhancing the qualities of heritage for user experience rather than preventing physical change [58–60].

9. Conclusions

The results of the study shed light on the evolving conservation practices at the Humble Administrator's Garden (HAG), exemplifying the complexities and challenges faced in preserving cultural heritage within the context of China's tourism-driven approach. Our investigation combined both historical and firsthand data to dissect the HAG's conservation evolution. We have evidenced how this has led to a somewhat predictable approach toward conservation methods and practices, with an initial focus on the faithful restoration of the spatial form of both architecture and landscape. The Li Residence has been transformed from a garden museum to a residential showcase space as part of the garden's prioritization as the main tourist resource. Yet this transformation—driven by the need to manage high tourist traffic—raises questions about the balance between architectural preservation and intangible cultural heritage representation within the site.

The growing global interest in more intangible and integrated approaches towards heritage highlights how cultural heritage landscapes are rich in meanings and values, serving as environments of memory and accommodating various contemporary interpretations of 'place'. From this critical perspective, we have consolidated interviews and visitor journals to show how the HAG's immersive multimedia tour attempts to revive a fading sense of unity between architecture and landscape with reimagined intangible heritage—such as literati aesthetics, activities, and feelings—being translated into scenes within the Li Residence and garden. Previously silent narratives are actively shaped and foregrounded by encouraging personal experiences and interpretations of the site's physical and non-physical heritage. This creates opportunities for further reflections on the tacit knowledge, practices, complex feelings, and stories that constitute the HAG heritage landscape in perpetuating its relevance to contemporary society. It also distinguishes it from common practices amongst other WHS gardens. We have indicated how this approach also dilutes the rigid division between architecture and landscape, demonstrating the complex interplay between the two elements. The concept of 'place' subsequently becomes more inclusive, considering both natural and built heritage and both tangible and intangible heritage domains.

It is clear from archives and interview results that the involvement of various stakeholders at the HAG, including the HAG Management Office, local government, conservation practitioners, and visitors, emphasizes the contested nature of heritage commodification. The immersive multimedia tour, a commodity-led initiative, has proven successful in generating economic profit whilst also inviting diverse interpretations and emotions that necessitate a constant reappraisal of the conservation strategies employed. The study's mixed methodological approach highlights the importance of involving previously silent stakeholders in the conservation process, with the HAG Management Office's multimedia tour facilitating subtle negotiations within the government-market heritage management system.

Ultimately, the conservation of cultural heritage landscapes must strive for a more holistic approach, encompassing tangible and intangible elements and embracing diverse narratives and interpretations. The HAG's attempted experiment to reuse, reinterpret, and represent its former heritage buildings indicates a changing conservation discourse where intangible heritage is influencing approaches toward physical heritage assets. This is a critical step towards the formation of a holistic narration of place that can more appropriately conserve and evolve WHS gardens and alike for future generations.

Author Contributions: Conceptualization, Y.R. and J.D.; methodology, Y.R. and J.D.; validation, Y.R. and J.D.; formal analysis, Y.R.; investigation, Y.R. and J.D.; resources, Y.R. and J.D.; data curation, Y.R.; writing—original draft preparation, Y.R. and J.D.; writing—review and editing, J.D.; visualization, Y.R.; project administration, J.D. All authors have read and agreed to the published version of the manuscript. We thank the Humble Administrator's Garden Management Office for their kind support.

Funding: This research received no external funding.

Institutional Review Board Statement: Consent forms and letters of approval to publish were acquired from the HAG Management Office prior to the production of this research article.

Informed Consent Statement: Informed consent was obtained from all subjects involved in the study.

Data Availability Statement: No new data were created or analyzed in this study. Data sharing is not applicable to this article.

Conflicts of Interest: The authors declare no conflict of interest.

References

1. Herbert, D. Literary places, tourism and the heritage experience. *Ann. Tour. Res.* **2001**, *28*, 312–333. [CrossRef]
2. Leask, A.; Fyall, A. *Managing World Heritage Sites*; Routledge: Oxfordshire, UK, 2006. [CrossRef]
3. Li, Y.; Lau, C.; Su, P. Heritage tourism stakeholder conflict: A case of a World Heritage Site in China. *J. Tour. Cult. Chang.* **2020**, *18*, 267–287. [CrossRef]
4. Li, M.; Wu, B.; Cai, L. Tourism development of World Heritage Sites in China: A geographic perspective. *Tour. Manag.* **2008**, *29*, 308–319. [CrossRef]

5.   Yang, C.-H.; Lin, H.-L.; Han, C.-C. Analysis of international tourist arrivals in China: The role of World Heritage Sites. *Tour. Manag.* **2010**, *31*, 827–837. [CrossRef]

6.   Zhu, Y. Cultural effects of authenticity: Contested heritage practices in China. *Int. J. Herit. Stud.* **2014**, *21*, 594–608. [CrossRef]

7.   Cohen, E.; Cohen, S.A. Authentication: Hot and cool. *Ann. Tour. Res.* **2012**, *39*, 1295–1314. [CrossRef]

8.   Li, Y. Heritage Tourism: The Contradictions between Conservation and Change. *Tour. Hosp. Res.* **2003**, *4*, 247–261. [CrossRef]

9.   Zhang, R.; Wang, J.; Brown, S. 'The Charm of a Thousand Years': Exploring tourists' perspectives of the 'culture-nature value' of the Humble Administrator's Garden, Suzhou, China. *Landsc. Res.* **2021**, *46*, 1071–1088. [CrossRef]

10.   Henderson, R. *The Gardens of Suzhou*; Pennsylvania Press: Philadelphia, PA, USA, 2012.

11.   Labadi, S. *UNESCO, Cultural Heritage, and Outstanding Universal Value: Value-based Analyses of the World Heritage and In-Tangible Cultural Heritage Conventions*; Altamira Press: Plymouth, UK, 2013.

12.   Avrami, E.; Mason, R. Mapping the issue of values. In *Values in Heritage Management: Emerging Approaches and Research Directions*; Avrami, E., Macdonald, S., Mason, R., Myers, D., Eds.; The Getty Conservation Institute: Los Angeles, CA, USA, 2019; pp. 9–34.

13.   Smith, L. *Uses of Heritage*; Routledge: Oxfordshire, UK, 2006.

14.   UNSECO. *Convention Concerning the Protection of the World Cultural and Natural Heritage*; UNESCO: Paris, France, 1972.

15.   Bortolotto, C. From the 'Monumental' to the 'Living' Heritage: A Shift in Perspective. In *World Heritage. Global Challenges, Local Solutions*; Ironbridge Institute: Oxford, UK, 2006; pp. 39–45.

16.   ICOMOS. *Historic Gardens (The Florence Charter)*; ICOMOS: Florence, Italy, 1982.

17.   Harrison, R. *Heritage: Critical Approaches*; Routledge: Oxfordshire, UK, 2013.

18.   Viñas, S.M. Contemporary theory of conservation. *Stud. Conserv.* **2002**, *47*, 25–34. [CrossRef]

19.   Craith, M.N. Heritage politics and neglected traditions: A case-study of Skellig Michael. In *Heritage Regimes and the State*; Bendix, R.F., Eggert, A., Peselmann, A., Eds.; Universitätsverlag Göttingen: Göttingen, Germany, 2013; pp. 157–176.

20.   Jones, S. Making place, resisting displacement: Conflicting national and local identities in Scotland. In *The Politics of Heritage: The Legacies of "Race"*; Littler, J., Naidoo, R., Eds.; Routledge: Oxfordshire, UK, 2004; pp. 94–114.

21.   Smeets, R.; Deacon, H. The examination of nomination files under the UNESCO Convention for the Safeguarding of the Intangible Cultural Heritage. In *The Routledge Companion to Intangible Cultural Heritage*; Stefano, M.L., Davis, P., Eds.; Routledge: London, UK, 2016; pp. 22–39.

22.   UNESCO. *Convention for the Safeguarding of the Intangible Cultural Heritage*; UNESCO: Paris, France, 2003.

23.   Pocock, C.; Collett, D.; Baulch, L. Assessing stories before sites: Identifying the tangible from the intangible. *Int. J. Herit. Stud.* **2015**, *21*, 962–982. [CrossRef]

24.   Su, J. Conceptualising the subjective authenticity of intangible cultural heritage. *Int. J. Herit. Stud.* **2018**, *24*, 919–937. [CrossRef]

25.   Silverman, H. Heritage and authenticity. In *The Palgrave Handbook of Contemporary Heritage Research*; Waterton, E., Watson, S., Eds.; Palgrave Macmillon: Hampshire, UK, 2015; pp. 69–88.

26.   Lenzerini, F. Intangible Cultural Heritage: The Living Culture of Peoples. *Eur. J. Int. Law* **2011**, *22*, 101–120. [CrossRef]

27.   Bortolotto, C. The French inventory of intangible cultural heritage: Domesticating a global paradigm into French heritage regime. In *Heritage Regimes and the State*, 2nd ed.; Bendix, R.F., Eggert, A., Peselmann, A., Eds.; Universitätsverlag Göttingen: Göttingen, Germany, 2013; pp. 265–282.

28.   Skounti, A. The authentic illusion: Humanity's intangible cultural heritage, the Moroccan experience. In *Intangible Heritage (Key Issues in Cultural Heritage)*; Smith, L., Akagawa, N., Eds.; Routledge: Oxfordshire, UK, 2009; pp. 74–92.

29.   Petronela, T. The Importance of the Intangible Cultural Heritage in the Economy. *Procedia Econ. Financ.* **2016**, *39*, 731–736. [CrossRef]

30.   Caust, J.; Vecco, M. Is UNESCO World Heritage recognition a blessing or burden? Evidence from developing Asian countries. *J. Cult. Herit.* **2017**, *27*, 1–9. [CrossRef]

31.   Hill, M.J. World Heritage and the Ontological Turn: New Materialities and the Enactment of Collective Pasts. *Anthr. Q.* **2018**, *91*, 1179–1202. [CrossRef]

32.   ICOMOS. *The Burra Charter: The Australia ICOMOS Charter for Places of Cultural Significance*; Australia ICOMOS: Burwood, Australia, 2013.

33.   Stubbs, J.H.; Makaš, E.G. *Architectural Conservation in Europe and the Americas*; John Wiley & Sons: New Jersey, NJ, USA, 2011.

34.   Harvey, D.C. Heritage Pasts and Heritage Presents: Temporality, meaning and the scope of heritage studies. *Int. J. Herit. Stud.* **2001**, *7*, 319–338. [CrossRef]

35.   Waterton, E. *Politics, Policy and the Discourses of Heritage in Britain*; Springer: Berlin, Germany, 2010. [CrossRef]

36.   Dahlin, J. Labour of love and devotion? The search for the lost soldiers of Russia. In *Emotion, Affective Practices, and the Past in the Present*; Smith, L., Wetherell, M., Campbell, G., Eds.; Routledge: London, UK, 2018; pp. 25–38.

37.   Edensor, T. National Identity and the Politics of Memory: Remembering Bruce and Wallace in Symbolic Space. *Environ. Plan. D Soc. Space* **1997**, *15*, 175–194. [CrossRef]

38.   Carr, G. Examining the memorialscape of occupation and liberation: A case study from the Channel Islands. *Int. J. Herit. Stud.* **2012**, *18*, 174–193. [CrossRef]

39.   Scher, P.W. Uneasy heritage: Ambivalence and ambiguity in Caribbean heritage practices. In *Heritage Regimes and the State*; Bendix, R.F., Eggert, A., Peselmann, A., Eds.; Universitätsverlag Göttingen: Göttingen, Germany, 2013; pp. 79–96.

40. Graham, B.; Ashworth, G.J.; Tunbridge, J.E. *A Geography of Heritage: Power, Culture, and Economy*; Routledge: Oxfordshire, UK, 2016.
41. Shea, M. Troubling heritage: Intimate pasts and troubling memories at Derry/Londonderry's 'temple. In *Emotion, Affective Practices, and the Past in the Present*; Smith, L., Wetherell, M., Campbell, G., Eds.; Routledge: London, UK, 2018; pp. 39–55.
42. Jokilehto, J. *A History of Architectural Conservation*, 2nd ed.; Routledge: London, UK, 2018.
43. Djabarouti, J. Practice barriers towards intangible heritage within the UK built heritage sector. *Int. J. Herit. Stud.* **2021**, *27*, 1101–1116. [CrossRef]
44. Djabarouti, J. Imitation and intangibility: Postmodern perspectives on restoration and authenticity at the Hill House Box, Scotland. *Int. J. Herit. Stud.* **2021**, *28*, 109–126. [CrossRef]
45. Yarrow, T. How conservation matters: Ethnographic explorations of historic building renovation. *J. Mater. Cult.* **2018**, *24*, 3–21. [CrossRef]
46. Connerton, P. *How Societies Remember*; Cambridge University Press: Cambridge, UK, 1989.
47. Misztal, B. *Theories of Social Remembering*; Open University Press: Berkshire, CA, USA, 2003.
48. Ivankova, N.V.; Creswell, J.W.; Stick, S.L. Using Mixed-Methods Sequential Explanatory Design: From Theory to Practice. *Field Methods* **2006**, *18*, 3–20. [CrossRef]
49. Yin, R.K. *Qualitative Research from Start to Finish*; The Guilford Press: New York, NY, USA, 2011.
50. Yi, X. The Tangible and Intangible Value of the Suzhou Classical Gardens. In Proceedings of the 16th ICOMOS General Assembly and International Symposium 'Finding the Spirit of Place—Between the Tangible and the Intangible', Quebec, QC, Canada, 29 September–4 October 2008.
51. Nash, C. Performativity in practice: Some recent work in cultural geography. *Prog. Hum. Geogr.* **2000**, *24*, 653–664. [CrossRef]
52. Knudsen, B.T.; Anne, M.W. (Eds.) Performative authenticity in tourism and spatial experience: Rethinking the relations between travel, place and emotion. In *Re-Investing Authenticity: Tourism, Place and Emotions*; Channel View Publications: Bristol, UK, 2010; pp. 1–20.
53. Zhang, H.Q.; Chong, K.; Ap, J. An analysis of tourism policy development in modern China. *Tour. Manag.* **1999**, *20*, 471–485. [CrossRef]
54. Cameron, E. New geographies of story and storytelling. *Prog. Hum. Geogr.* **2012**, *36*, 573–592. [CrossRef]
55. Hodder, I. The Entanglements of Humans and Things: A Long-Term View. *New Lit. Hist.* **2014**, *45*, 19–36. [CrossRef]
56. Djabarouti, J. Stories of feelings and things: Intangible heritage from within the built heritage paradigm in the UK. *Int. J. Herit. Stud.* **2020**, *27*, 391–406. [CrossRef]
57. Walter, N. From values to narrative: A new foundation for the conservation of historic buildings. *Int. J. Herit. Stud.* **2013**, *20*, 634–650. [CrossRef]
58. Waterton, E. Archaeology, visuality and the negotiation of heritage. In *Taking Archaeology Out of Heritage*; Waterton, E., Smith, L., Eds.; Cambridge Scholars Publishing: Newcastle Upon Tyne, UK, 2009; pp. 28–47.
59. Zetterstrom-Sharp, J. Heritage as future-making: Aspiration and common destiny in Sierra Leone. *Int. J. Herit. Stud.* **2014**, *21*, 609–627. [CrossRef]
60. Fredheim, L.H.; Khalaf, M. The significance of values: Heritage value typologies re-examined. *Int. J. Herit. Stud.* **2016**, *22*, 466–481. [CrossRef]

Disclaimer/Publisher's Note: The statements, opinions and data contained in all publications are solely those of the individual author(s) and contributor(s) and not of MDPI and/or the editor(s). MDPI and/or the editor(s) disclaim responsibility for any injury to people or property resulting from any ideas, methods, instructions or products referred to in the content.

Article

Lost in Translation: Tangible and Non-Tangible in Conservation

Nigel Walter

Department of Archaeology, University of York, York YO1 7EP, UK; nhw502@york.ac.uk

Abstract: This paper addresses the special issue theme of the response of conservation practice to shifts in heritage theory towards the intangible, through exploring some specific aspects of practice and statutory process in the UK. The paper starts with an overview of conservation in the UK, and the extent to which it does or does not interface with developments in heritage theory. It explores the conventional understanding of significance—here termed 'subtractive'—which reflects the antiquarian concerns from which conservation developed. It then considers the Ecclesiastical Exemption, a parallel consent mechanism within UK law for Christian places of worship that remain in use, which specifically recognises their need to change over time to ensure their survival. Evidence for a growing appreciation of non-tangible value and community participation in heritage is provided in recent research by The National Churches Trust into the economic and social value of church buildings to local communities across the UK. The paper concludes that a positive response to changes in heritage theory requires conservation to undertake its own theoretical work; this will involve a recognition of living buildings as central rather than peripheral both to conservation and to heritage more broadly, and a move towards a 'generative' understanding of significance.

Keywords: conservation; intangible heritage; Ecclesiastical Exemption; Faculty Jurisdiction; significance; National Churches Trust; heritage theory

Citation: Walter, N. Lost in Translation: Tangible and Non-Tangible in Conservation. *Architecture* **2023**, *3*, 578–592. https://doi.org/10.3390/architecture3030031

Academic Editor: Johnathan Djabarouti

Received: 25 June 2023
Revised: 2 August 2023
Accepted: 7 August 2023
Published: 21 September 2023

Copyright: © 2023 by the author. Licensee MDPI, Basel, Switzerland. This article is an open access article distributed under the terms and conditions of the Creative Commons Attribution (CC BY) license (https://creativecommons.org/licenses/by/4.0/).

1. Introduction

This paper addresses the special issue theme of the response of conservation practice to the shift of heritage theory from its former focus on tangible product towards an understanding of intangible process. It does this through exploring some specific aspects of practical experience of the conservation of living buildings in the UK, particularly historic church buildings under the Ecclesiastical Exemption; the parallel issue of the adaptive reuse of historic church buildings no longer in use raises a complementary but distinct set of issues and is not addressed.

For the purposes of the paper, tangible heritage can be defined as material forms of heritage, including historic buildings, monuments, artefacts, artworks, archaeological remains, etc. The labelling of these as 'tangible' flowed from the recognition in the later twentieth century of other forms of heritage that are 'intangible' [1–3], a shift that culminated in the 2003 Convention for the Safeguarding of the Intangible Cultural Heritage, which defines intangible cultural heritage as 'the practices, representations, expressions, knowledge, skills [...] that communities, groups and, in some cases, individuals recognize as part of their cultural heritage' ([3], art. 2.1). This is understood to be 'transmitted from generation to generation', and is 'constantly recreated' by communities and groups, moving the focus of heritage away from a physical product towards an intergenerational process.

The nature of the interrelation of these two forms of heritage has been a cause of concern from the outset; in 2004, the *Yamato Declaration on Integrated Approaches for Safeguarding Tangible and Intangible Cultural Heritage* noted their interdependence and called for the elaboration of 'integrated approaches', without exploring the nature of their interrelation [4]. A few weeks later, the World Heritage Committee noted the relevance of the *Declaration* but stressed that the two conventions 'address different forms of heritage and therefore [. . .] have different scopes' [5]. The resulting understanding—at least in a World Heritage context—is of two domains of heritage which overlap, but which remain fundamentally dissimilar.

This paper explores this relationship between the tangible and, using the earlier and now more neutral term, the 'non-tangible'; it aims to question the conventional distinction between the two, arguing instead for continuity between them.

2. Conservation Practice in England and Beyond

In a UK context, Historic England is the national government's statutory adviser on the historic environment in England; central amongst its varied range of guidance for heritage professionals is its 2008 document *Conservation Principles, Policies and Guidance* [6]. Notwithstanding a controversial and ill-fated attempt at revision in 2018, this remains the organisation's core guidance document [1]. In methodology, it adapts the *Burra Charter's* fourfold structure of values, and adopts the language of people and place [7]. It defines conservation as:

> the process of managing change to a significant place in its setting in ways that will best sustain its heritage values, while recognising opportunities to reveal or reinforce those values for present and future generations ([6], p. 7).

This placing of the 'management of change' at the centre of conservation has far-reaching consequences, and remains a significant issue for some heritage professionals, because it challenges the previously dominant preservationist understanding, that responsible conservation necessarily involves constraining change to the minimum. By contrast, Historic England declares an openness to change, for example noting that:

> The concept of conservation area designation, with its requirement 'to preserve or enhance', also recognises the potential for beneficial change to significant places, to reveal and reinforce value ([6], p. 15).

Clearly, conservation areas are subject to a different form of protection from that of individual buildings; nevertheless, there is a clear interrelation, which *Conservation Principles* itself underlines, and which flows from that document's adoption of the *Burra Charter's* use of 'place' as a richer and more-than-tangible term than 'building', 'site', 'monument', etc. ([6], e.g., pp. 14–15, 38). Similarly, the UK's overarching *National Planning Policy Framework* (*NPPF*) speaks of 'the desirability of sustaining and *enhancing* the significance of heritage assets' ([8], emphasis added).

This question of how change to historic buildings should be regarded has been a constant concern for conservation from its inception. Under his ICOMOS presidency from 2008 to 2016, Gustavo Araoz spurred an international debate over a 'New Heritage Paradigm', central to which is what he termed the historic environment's 'tolerance for change'. In a paper delivered to the ICOMOS Advisory Committee in Valletta in October 2009, he characterised the development of conservation thus:

> During the 19th and most of the 20th century, the heritage conservation community developed under the assumption that all values attributed to places rested on the material evidence of the place. Thus, the theory and praxis of conservation evolved [...] as an increasingly sophisticated effort *to prevent form and space from undergoing changes* ([9], emphasis added).

This argument was played out in two ICOMOS conferences in 2010 and 2011, including a version of Araoz's paper and a robust response from his predecessor as President, Michael Petzet [10,11]. Petzet was particularly keen to challenge the idea of conservation as the management of change, describing this as an 'inconsiderate [unconsidered?] general proposal', and warning that:

> the core ideology of our organization is being counteracted. After all, conservation does not mean 'managing change' but preserving—preserving, not altering and destroying: ICOMOS [. . .] is certainly not an International Council on Managing Change [12,13].

Petzet also took critical aim at the *Burra Charter*, controversially dismissing this as 'an "Australian" heritage philosophy [which] is quite confusing and suitable for damaging the traditional objectives of monument conservation' ([13], pp. 10–11).

That debate overlapped with the development of the *Recommendation on the Historic Urban Landscape* (HUL) [14], and Araoz specifically cites historic urban areas in support of his argument: '...an important cultural value of the historic city rests precisely upon its ability to be in a constant evolution ...' ([9], p. 58). HUL, interestingly, is an aspect of international conservation almost entirely ignored by Historic England [15] [2].

From this selective overview, it is clear that the approach to change in conservation can be said to be significantly contested, as much internationally as in the UK. On the one hand, Historic England's overarching guidance borrows and develops the *Burra Charter's* fourfold values structure, according the non-tangible (in the form of communal value) a central role in the heritage process. On the other hand, the implications of this broadening of heritage are by no means universally accepted, with much of the discourse around heritage continuing to be argued from a tangible-centric and preservationist position.

The argument between Petzet and Araoz is at first sight perplexing. How can two such prominent heritage professionals—successive presidents of the leading international conservation body, no less—espouse such different approaches? To begin to account for this, it is essential to acknowledge that conservation does not operate only at the level of the explicit knowledge contained in closely worded conservation charters, protocols, and guidance—which together form the discipline's 'Doctrinal Texts'. Like any human endeavour, conservation is also replete with *implicit* knowledge—the underlying structure that scientist and philosopher Michael Polanyi termed the 'tacit dimension' [16]. This feeds and shapes our understanding, and thus the knowledge we articulate, and the processes by which we organise our discipline. Noting the role played by this 'tacit dimension' is by no means a criticism; rather, it shows that conservation is a living tradition, that is, an ongoing, intergenerational, communal argument over what constitutes the good [17]. However, this brings into play such fundamental questions as our understanding of the purpose of the discipline—an explicit concern for Petzet—and what good practice looks like. It is at that foundational level that the course of conservation's original development remains pertinent and active.

Both individual buildings, and conservation as a whole, are contested. In 1906, the novelist Thomas Hardy (who, before becoming a writer, had himself been an architect) presented a paper to the SPAB annual meeting, entitled 'Memories of Church Restoration' [18]. Hardy stresses the contested character of historic churches, and merits quoting at length:

> At first sight it seems an easy matter to preserve an old building without hurting its character. Let nobody form an opinion on that point who has never had an old building to preserve.
>
> In respect of church conservation, the difficulty we encounter on the threshold, and one which besets us at every turn, is the fact that the building is beheld in two contradictory lights, and required for two incompatible purposes. To the incumbent the church is a workshop or laboratory; to the antiquary it is a relic. To the parish it is a utility; to the outsider a luxury. How [to] unite these incompatibles? A utilitarian machine has naturally to be kept going, so that it may continue to discharge its original functions; an antiquarian specimen has to be preserved without making good even its worst deficiencies. The quaintly carved seat that a touch will damage has to be sat in, the frameless doors with the queer old locks and hinges have to keep out draughts, the bells whose shaking endangers the graceful steeple have to be rung.
>
> If the ruinous church could be enclosed in a crystal palace, covering it to the weathercock from rain and wind, and a new church be built alongside for services (assuming the parish to retain sufficient earnest-mindedness to desire them), the method would be an ideal one. But even a parish entirely composed of opulent

members of this Society would be staggered by such an undertaking. No: all that can be done is of the nature of compromise ([18], pp. 204–205).

Hardy is clear that this is a difficult issue. He characterises historic churches as contested between incumbent, antiquary, parish, and outsider. This plurality boils down to viewing the building 'in two contradictory lights', as something 'required for two incompatible purposes'. He labels these two purposes 'utilitarian machine' and 'antiquarian specimen', respectively, presaging the contemporary distinction between intangible and tangible, and firmly placing them in mutual opposition. His 'crystal palace' ideal is preservationism in its purest (and purist) form—the museumification of tangible heritage by literally placing the historic building in a glass box. Clearly, such an approach would be unfeasibly expensive, something Hardy immediately acknowledges. However, even were it feasible, it would be far from 'ideal'; indeed, I suggest it would be disastrous for the heritage such a building constitutes, holistically understood.

It is worth noting that this statement of the preservationist ideal may not only suit the latter-day antiquarian resolutely focused on the tangible, but in today's context potentially also suits those who see heritage as essentially intangible. Why? Because both ignore the interconnection of people and place. Both also tend to ignore the recognised importance of retaining buildings in beneficial use ([6], p. 43). Without an adequate account of how tangible and non-tangible are integrated, it is easy to argue that a community would be better able to express their intangible heritage if 'freed' of the 'burden' of an awkward historic building. This is precisely what Hardy, in older language, offers as his ideal, and also what some current church communities indeed long for; both, in my view, profoundly misread the nature and importance of tangible heritage.

Hardy's essay remains highly relevant, not only for his diagnosis of the contested nature of heritage, nor only for underlining the link between antiquarianism and preservation, nor again only for foregrounding the tension between the tangible/'antiquarian' and intangible/'utilitarian'. In addition to all of these, the relevance of Hardy's essay lies in demonstrating the continued influence of the preservationist approach as an animating force underpinning the mindset of many conservation practitioners.

From its antiquarian foundation, conservation has developed an art historical approach to historic buildings, which arguably still predominates. Clearly, it would not be possible for conservation practitioners to function responsibly without attending to the art historical aspects of a given building—without the art historical, one would be unable to situate the building in its context, to read its development, or to judge the building's capacity for change. But, however *necessary*, we should not fool ourselves that art history is, of itself, *sufficient*. We must recognise that all art history is written from a particular view, with a particular agenda—as an example, Zachary Stewart provides an excellent demonstration of this in a recent paper on the differing readings of the development of the Perpendicular style from the late fourteenth century onwards [19].

Not only can an art historical approach never be neutral, it inevitably brings with it a tendency to read historic buildings as completed artworks. The philosopher Hans-Georg Gadamer addresses this very issue, insisting that:

> A building is never only a work of art. Its purpose, through which it belongs in the context of life, cannot be separated from it without its losing some of its reality. If it has become merely an object of aesthetic consciousness, then it has merely a shadowy reality and lives a distorted life only in the degenerate form of a tourist attraction or a subject for photography. The "work of art in itself" proves to be a pure abstraction [20].

The abstraction of the 'work of art in itself' flies in the face of the reality of most historic buildings, which typically have changed multiple times over their history to date, and which, through their continuing purpose, stubbornly continue to 'belong in the context of life'. Some 30 years ago, Stewart Brand persuasively made the functional case that it is in the nature of buildings to change with his classic book *How Buildings Learn: What Happens After*

They're Built [21]. Furthermore, that history of change becomes an integral and inseparable part of their character, such that when we encounter historic buildings, it is often precisely for that unfolding, messy, and hybrid nature that we love them. Many historic buildings owe their very survival to that ability to adapt to changing needs; without changing, many of these buildings would have been discarded. Furthermore, their continuing use—whether for their core purpose or following adaptive reuse—will lead to further pressure for change. This is not to be regretted, but instead is a sign of health, of a strong link between people and place. Indeed, Historic England recognises this reality, stating that 'Keeping a significant place in use is likely to require continual adaptation and change' ([6], p.43).

3. Subtractive Significance and Critical Approaches

By contrast, approaching historic buildings as completed artworks effectively forecloses change. In the art world, conservationists rightly agonise over whether and how to restore an old artwork, and there are examples of parallel discussions in building conservation [22,23]. Few would credibly think they could propose elective change to a Rembrandt without compromising or destroying its integrity, its authenticity, its value (though this is precisely the issue Chinese artist Ai Weiwei has explored with his use of 2000-year-old Chinese vases, in self-descriptive works such as *Dropping a Han Dynasty Urn* (1995), and *Han Jar Overpainted with Coca-Cola Logo* (1995) [24,25]). We can describe this building-as-artwork model as displaying a 'subtractive' understanding of significance—that the work starts from a more or less maximal state of significance, and change can only reduce that bank of significance.

This view of significance as subtractive can be traced from the earliest days of conservation and remains dominant among many conservation professionals to this day. It is evident in the words used in official heritage writing; for example, the preambles to the World Heritage Convention has much to say about threat, harm, and loss:

> Noting that the cultural heritage and the natural heritage are increasingly *threatened* with *destruction* not only by the traditional causes of *decay*, but also by changing social and economic conditions which aggravate the situation with even more formidable phenomena of *damage* or *destruction,*

> Considering that *deterioration* or *disappearance* of any item of the cultural or natural heritage constitutes a *harmful impoverishment* of the heritage of all the nations of the world [. . .]

> Considering [. . .] the importance, for all the peoples of the world, of safeguarding this *unique* and *irreplaceable* property. . . ([26], emphasis added).

Familiar as it may be, the subtractive understanding tends towards absurdity when applied to the less-than-complete loss that results from change to historic buildings in use. If change can only harm significance, then a building such as a medieval church which will often have undergone a dozen or more episodes of change—at times including the removal of whole sections alongside alterations and additions—must have precious little significance left. Clearly that is not the case, because we more often value these buildings precisely *because of* those episodes of change; rather than lamenting each particular loss, we see benefit in the whole.

There must, therefore, be more going on than mere subtraction. Where previous episodes of change to these buildings were 'successful', it was because they took place within a tradition. William Morris recognised this in his celebrated 1877 manifesto for the Society for the Protection of Ancient Buildings, to which modern conservation in the UK generally dates its inception:

> a church of the eleventh century might be added to or altered in the twelfth, thirteenth, fourteenth, fifteenth, sixteenth, or even the seventeenth or eighteenth centuries; but every change, whatever history it destroyed, left history in the gap, and was alive with the spirit of the deeds done midst its fashioning [27].

Morris's claim is that such change took place in a bygone and irretrievable age. He regards historic buildings 'as monuments of a bygone art, created by bygone manners', pleading with us 'to remember how much is gone of the religion, thought and manners of time past' [27].

In this way, Morris, the great medievalist and champion of ancient buildings, declares himself a thoroughgoing modern. In a more critical mode, anthropologist Bruno Latour explicitly links this understanding of modernity—as following a definitive rupture with the past—with conservation:

> As Nietzsche observed long ago, the moderns suffer from the illness of historicism. They want to keep everything, date everything, because they think they have definitively broken with their past. The more they accumulate revolutions, the more they save; the more they capitalize, the more they put on display in museums. Maniacal destruction is counterbalanced by an equally maniacal conservation [28].

The commitment to modernity and its claimed break with the past—clearly evident in both Morris and Hardy after him—lies at the heart of the preservationist approach. How ironic, then, that we think it legitimate to entrust historic buildings (many dating from before the dawn of modernity) to a system that not only was developed in response to the excesses of modernity but—if Latour is right—is itself a direct expression of those excesses.

In several papers, Cornelius Holtorf has critiqued the 'conservation paradigm' which sees heritage as a finite and diminishing resource that must be defended at all costs, suggesting, provocatively, that 'destruction and loss are not the opposite of heritage but constitutive of it' [29–31]. In his 2015 paper, he applies the insights of loss aversion theory (from the field of economics) to cultural heritage, critiquing the longstanding preference in Western cultural heritage for avoiding loss over acquiring gains of the same value. He uses examples of wholesale loss—including the Fantoft Stave Church, Norway (destroyed by arson in 1992), the Gloucester home of murderers Fred and Rose West (demolished by the authorities), and Ai Weiwei's destruction of ancient Chinese vases noted above—to show that the significance of heritage can persist and indeed grow after their destruction. Holtorf's examples, and others like them, flatly contradict the subtractive understanding of significance.

Similarly, Rodney Harrison describes how heritage practice in the West at least up until the late twentieth century focused on the collection of remarkable buildings, terming this 'a *canonical* model of heritage' ([32], emphasis original). On this view, heritage is separated from and deemed more valuable than the present and its cultural production, and contrasts with more recent models he characterises as 'continuous'. As Harrison discusses, the adoption by conservation of the idea of the cultural landscape in response to the concerns of Indigenous peoples speaks powerfully of the interrelation of tangible and non-tangible ([32], pp. 114–139). As Graham Fairclough points out, the idea also has profound implications for the understanding of continuity and change:

> The idea of cultural landscape has the concept of change (in the future as well as in the past) at its very heart. The idea that there are any landscapes where time has stood still, and history has ended, is very strange. No landscape, whether urban or rural, has stopped its evolution, no landscape is relict: it is all continuing and ongoing [...]. The decision that each generation, including archaeologists has to make, is what will happen next to the landscape, and how it will be managed or changed [33].

The development in the last two decades of Historic Urban Landscape thinking, from the 2005 Vienna Memorandum to the 2011 Recommendations and beyond [14,34,35], has brought with it a holistic approach and a blurring of the boundaries between building, context and urban area [36]. As suggested by Gustavo Araoz's engagement with HUL (noted above), we can in time expect insights from the field of cultural landscape to inform conservation, to its enrichment.

That said, with the notable exception of Araoz, none of these mentioned thinkers who are challenging the preservationist 'conservation paradigm' comes from amongst practising conservation professionals. Indeed, most are archaeologists by background. The architects and surveyors who become conservation specialists often do so without further academic qualification, and even if they do postgraduate qualifications, cannot be guaranteed to encounter heritage studies thinking of the sort represented by Holtorf, Harrison, and Fairclough [37] [3].

4. Church Buildings

We turn next to an oddity of British heritage protection, rooted partly in pre-modernity, that is suggestive of a different understanding of the relation of tangible and non-tangible heritage.

The Ecclesiastical Exemption offers a parallel consent mechanism within UK law for Christian places of worship that remain in use; while not unique, this is unusual in international terms [4]. The Exemption is an administrative umbrella, a set of criteria within which individual Christian denominations can develop their own systems to control alterations to listed buildings. It is an exemption from listed building consent only, not from planning permission, nor from Scheduled Monument Consent, nor indeed from Building Regulations approval. The Exemption applies in different forms in each of the four nations of the UK; in England, which has the greatest number of protected buildings, the Exemption covers five denominations—the Church of England, the Roman Catholic Church, the Methodist Church, the United Reformed Church, and churches within the Baptist Union.

In 2010, the UK's Department of Culture, Media, and Sport (DCMS) published guidance to accompany the last revision to the legislation [38]. This guidance insists that any procedures under the Exemption 'must be as stringent as the procedures required under the secular heritage protection system'; this 'equivalence of protection' is identified as a key principle, and one which will be kept under review to ensure that appropriate standards of protection are maintained ([38], p. 7). There is always, therefore, the potential for the Exemption to be withdrawn, or indeed for a denomination itself to withdraw from it, as did the United Reformed Church in Wales in 2018.

The guidance appreciates the importance of keeping historic buildings in use, if they are to survive. It states:

> The Ecclesiastical Exemption reduces burdens on the planning system while maintaining an appropriate level of protection and reflecting the particular need of listed buildings in use as places of worship to be able to adapt to changing needs over time to ensure their survival in their intended use ([38], p.6).

This offers explicit recognition that, for living listed buildings such as churches, change is legitimate in principle, and essential to their survival as places of worship; at the same time, it points towards the brutal reality that, should the Exemption be withdrawn, the secular planning system would simply be unable to cope.

The Church of England has by far the greatest number of listed churches (some 12,300), of which 4300 are listed grade I and a further 4300 grade II*. Because of both the numbers and the highly listed nature of the buildings in question, the Church of England's Faculty Jurisdiction is the most developed system under the Exemption. The system dates from 1913, in the early days of heritage protection in England, and was built on the Church's existing faculty system, which dates back to medieval times, arguably, therefore, helpfully bringing with it elements of pre-modernity. As a result of this genesis, the system is legal in nature, with each diocese having a 'consistory court', presided over by a judge, known as the 'chancellor'. Contested decisions are set down in publicly available judgments [5]; these record the arguments in the case, and combine to form an invaluable research resource, as well as helping satisfy the Exemption requirement for openness and transparency.

Each diocese is required to maintain a Diocesan Advisory Committee (DAC), which should cover specific areas of expertise, including in the development and use of church buildings, in liturgy and worship, in architecture and archaeology, and in the care of historic buildings and their contents. DACs (and equivalent bodies in other exempt denominations) thus bring together a far wider range of specialist skills than the secular system has at its disposal, offering a vital source of pro bono expert advice.

While it goes to great lengths to mirror the rigour of the secular process, and to satisfy the requirement for openness and transparency, it is freely acknowledged that the Faculty Jurisdiction will often produce different results from the secular system. Because of this, some UK conservation professionals regard the Exemption as an aberration, giving unwarranted and unwelcome licence to the major church denominations to make their own rules, leading to the degrading and potential destruction of some of the nation's most important heritage. However, for the Exemption's defenders, the secular system is both ill-equipped and incapable of dealing adequately with the particularity of churches; on that latter view, a separate system is essential because the secular system does not understand the non-tangible aspects of heritage inherent in a church building. How, for example, could a secular conservation officer begin to assess the justification for liturgical changes such as a reordering of furniture, the introduction of a nave altar or the moving of a baptismal font, without themselves being conversant with such buildings in use?

One landmark decision is the 2012 appeal judgment over St Alkmund, Duffield, which now plays a central role in the Faculty system, providing an established framework for guiding chancellors in their judgments, and acting as a common point of reference for all parties ([39]; for a fuller discussion of this judgment, see [40]). It also powerfully illustrates how non-expert voices can be closely attended to alongside the expert, providing evidence of a welcome rebalancing of that sometimes-fraught relationship. The Duffield judgment, along with many others like it, suggests that the Exemption is an example of an official conservation process that recognises the reality of living heritage. While it may be unusual, the Exemption is thus highly significant—perhaps as some form or prototype or forerunner—in the present context of the shift in heritage theory. Further, the Exemption demonstrates that this shift need not be *away from* the tangible, but instead *towards* a holistic understanding of heritage as a nexus—literally a binding together—of tangible and non-tangible.

A second demonstration of a growing appreciation of non-tangible value and community participation in heritage is offered by some recent research by the National Churches Trust (NCT), a grant-making and campaigning charity; that research aimed to quantify the economic and social value of church buildings to the UK [41] [6]. The resulting *House of Good* report is based on HM Treasury's official Green Book methodology—the government's own means of assessing social value—and concludes that the annual contribution of church buildings to the UK economy is an extraordinary 55 billion GBP. Given that there are estimated to be just under 40,000 places of Christian worship in use in the UK, that is an average of 1.4 million GBP for each and every church building. Over three quarters of that astounding total lies in the wellbeing value to individuals from community activities run or hosted by churches, such as food banks, youth groups, drug and alcohol support, etc.

Naturally, not all of those 40,000 churches are buildings of tangible heritage significance, though perhaps 40% are, including, as we have seen, just under half of all of England's grade I listed buildings. Furthermore, the sorts of activities that attract such high levels of social value in the report take place in buildings across the full range of heritage value.

Clearly, this sort of financial methodology—designed to help central government compare policy priorities—can never deliver a complete description of the rich significance of historic church buildings. However, while no one would claim the description to be all-encompassing, it surely does form a necessary part of any balanced assessment. Heritage voices from a conservative position that might argue that food banks, etc., are at best incidental to the life of a historic church building miss something vital (literally) about living buildings: that such activities represent a continuation of a tradition that has been a part of this heritage—the amalgam of people and place—from their inception. Furthermore, these non-religious, social activities are also highly relevant to the question at hand; such research demonstrates plainly that there are categories other than the purely material that must be considered when determining the importance of historic buildings in use.

5. The Centrality of Living Buildings

The startling figures in the NCT research raise the issue of 'living buildings', what is meant by that evocative term, and what implications this may have for conservation as a whole. The distinction of 'living' from 'dead' buildings dates from the early days of heritage protection, for example featuring in the first three of six resolutions of the Sixth International Congress of Architects held in Madrid in 1904. The first reads:

> Monuments may be divided into two classes, *dead monuments*, i.e. those belonging to a past civilization or serving obsolete purposes, and *living monuments*, i.e. those which continue to serve the purposes for which they were originally intended ([42], emphasis original).

The second resolution states that 'Dead monuments should be preserved . . .', while the third says that 'Living monuments ought to be restored so that they may continue to be of use, for in architecture utility is one of the bases of beauty.' It is the fact that these buildings continue in use that is seen as so pivotal, and the last comment on the relation of beauty and utility is commensurate with Gadamer's view of buildings discussed in Section 2 above.

Church buildings can be termed 'living' in at least three, typically overlapping, senses: first, in almost all cases, these buildings continue in the use for which they were first built; second, their principal users are a community, previous generations of which generally created the building in the first place and whose continuous presence has been at the heart of the building ever since; and, third, the building has typically changed multiple times through its history, thus being more dynamic than static in nature.

Veteran conservation architect Donald Insall chose *Living Buildings* for the title for a retrospective monograph covering 50 years of practice; he argues that all historic buildings are 'alive and constantly changing', whether through successive intervention or the effects of time and weather [43]. In an early chapter entitled 'Buildings are Alive', he offers the example of St Anne's Church, Kew which, since 1714, has evolved through nine distinct stages ([43], pp. 42–43). If, as Insall suggests, all buildings are alive in this way—at least to some degree—then this poses fundamental questions for conservation, and rules out any notion that the purpose of conservation might be 'to keep things the same'.

And yet, I suggest that for as long as the discipline clings to its preservationist roots—as exemplified in Holtorf's 'conservation paradigm' and Hardy's comment in Section 2—preservation will remain its basic orientation, feeding a residual bias against change, however judicious. This shows in the use of language, e.g., in the frequent use of the word 'harm' where 'change' or 'impact' would be more appropriate, as touched on in Section 3 above. Clearly, *some* change can be extremely harmful to heritage, as the discipline knows to its cost, but that is not true of *all* change. The view that all change to historic buildings is harmful is one that can be argued, but it is far from neutral, being heavily invested in that preservationist understanding of heritage discussed earlier.

The understanding that buildings grow and acquire identity as they do so is paralleled by social anthropologist Tim Ingold, known for his phenomenological approach to the relation of people to their material environment. Ingold questions the dominant 'genealogical model' within archaeology, 'namely that persons and things are virtually constituted, independently and in advance of their material instantiation in the lifeworld' [44]. This he roots in Aristotelian hylomorphism, the belief that physical entities are compounds of matter and immaterial form (for more of Ingold's critique of hylomorphism, see [45]). In place of this understanding of people and things as 'created in advance', he suggests it is instead better 'to think of a world not of finished entities, each of which can be attributed to a novel conception, but of processes that are *continually carrying on*, and of forms as the more or less durable envelopes or crystallisations of these processes' ([44], p. 163, emphasis original). He helpfully critiques traditional archaeology as 'too concerned with that which has been preserved in the archaeological record, which he characterises as fragments that have broken off from the flow of time, [and that] recede ever further from the horizon of the present. They become older and older, held fast to the moment, while the rest of the world moves on. But by the same token, the things of the archaeological record do not persist. For whatever persists carries on, advancing on the cusp of time ([44], p. 164).

In language very similar to that used by Gadamer above, this notion that only those things that advance 'on the cusp of time' persist is a radical restatement of the accepted conservation wisdom that continued beneficial use of historical buildings is desirable ([6], p. 43; see also [46]). The parallels with living buildings are clear; Ingold's restatement is highly relevant to living buildings.

If living buildings are moved from the status of being a special case to being seen as the norm—as Ingold's argument would imply, and Insall suggests they should—then this would have profound implications for conservation practice, particularly in the need to do the urgent work of developing a more robust theoretical foundation for the discipline. Important work has already been done to that end, including under the Living Heritage Sites Programme at the 'International Centre for the Study of the Preservation and Restoration of Cultural Property' (ICCROM). Ioannis Poulios, in his important study of the monastic sites at Meteora in Greece, has identified four forms of continuity that sustain living heritage: continuity of function, of relation between a core community and the tangible heritage, of expressions (entailing a recognition that heritage places will continue to change), and of care [47]. To be clear, continuity here refers to dynamic continuity of an ongoing process, not the static continuity of an unchanging product. And that third form of continuity, of expression, entails ongoing creativity, something Poulios foregrounded in an earlier paper [48]. A similar fourfold continuity is present in the excellent paper *Living Heritage: A Summary*, published by ICCROM [49].

We can note, therefore, that a living heritage approach is distinct from much conventional Western conservation thinking, in two respects of particular relevance to this discussion. First, a living heritage approach acknowledges the principle of the legitimacy of change. Second, just as the community across time needs to be considered as a whole—by placing it within the ongoing and intergenerational argument that is its tradition—so too the building itself should be valued for the open-ended holism that encompasses all its stages of development, not only for the current stage at which the story happens to have paused for a time. That is a temporal perspective that comes from operating within an active tradition. In the case of church buildings, this would be the Christian tradition, central to which is the idea of the Communion of Saints—that is, that the Church is a community extended through time, including those generations long past, those currently alive and, by extension, those still to come [50]. When dealing with buildings of that tradition, it is surely more appropriate to deal with them in this way, rather than as modernity does, as specimens of a bygone art, as we saw Morris suggesting above.

Conservation is faced with a choice—and one that will define it for generations to come—over whether or not to engage with living heritage, and allow itself to be changed by it. It will be very attractive to hunker down in the tangible realm, but this requires discounting the relevance of those community voices who, in attempting to live well with their buildings, seek to change them. Or, conservation can join Heritage Studies in attending to those community voices, embracing living buildings and welcoming the enrichment of the discipline that will result. The outcome of that positive turn would be more change to historic buildings, but also more historic buildings in better condition and with better prospects of surviving long into the future. And that, it should be stressed, should *not* be seen as a trade-off, tolerating change and even loss to achieve the benefit of survival, as with the Fantoft stave church and other extreme examples discussed above. Rather, a living buildings perspective puts *both* factors on the positive side of the balance.

For conservation to respond positively in the way described will require it to undertake its own theoretical work. In particular, this will involve a recognition of living buildings as central rather than peripheral both to conservation and to heritage more broadly. It will also require reflection on the nature of change, and from that the development of guidelines and processes for distinguishing change that causes unacceptable harm, from change that enriches a building by continuing its narrative. Conservation's current processes are not adequate to that task. They must be rethought, which will involve tracing through the implications of choosing a different starting point from that of the preservationists who so effectively set the agenda for the early stages of conservation's development.

6. 'Generative' Significance

What, then, are the implications for our understanding of significance? At the start of this paper, the current framing of significance was characterised as 'subtractive', based on a predominantly art historical understanding of built heritage. The Ecclesiastical Exemption, the *House of Good* report, the living building approach, and other heritage voices each in their different ways illustrate the limitations of such an understanding. This final section considers what an alternative understanding of significance might look like.

Given that this theoretical work is done in the context of recognising the creative and ongoing nature of living heritage, we could provisionally label this alternative view of significance as 'generative'. In practice, many conservation professionals rarely engage with theory, instead seeing the aesthetic, the historical, and the techniques of material conservation as the core of the discipline; this priority is reflected in the 14 criteria listed in the ICOMOS *Guidelines for Education and Training* [51]. And yet, this theoretical work is unavoidable. Even if we wanted to avoid theory, we cannot—one either engages with it deliberately, or one finds oneself manipulated by the theory of others.

It should be restated that this paper is not arguing against the discipline of architectural history or, more generally, art history. The art historical will be a prominent concern in most cases of physical heritage. We can say that it will always be a necessary part of conservation, but of itself it can never be sufficient. And that goes back to our definition of the heritage we are attempting to conserve—not place (tangible) alone, nor indeed people (intangible) alone, but a hybrid or an amalgam or a binding together of the two. The issue is that unless the foundational understanding that sees conservation in predominantly art historical terms is directly challenged, the subtractive understanding of significance will endure. From that source flows a conservation system built around resistance to change, at the attendant cost of much heritage being 'lost in translation'.

Under a 'generative' view of significance, by contrast, the identity of a historic building is not seen as something fixed, but as something that continues to develop. It remains possible that the building may be harmed by proposed change but, equally, with the right approach and craft skills, its significance can be enhanced. Such a view of significance would be a much better fit for the sort of living buildings described above, buildings with a long history of change. And in an English context, such a view fits well with government policy, which requires local planning authorities to take account of 'the *desirability* of sustaining and enhancing the significance of heritage assets and putting them to viable uses consistent with their conservation' [8].

The generative view also fits well with the *Burra Charter* (and, by extension, with parts at least of heritage policy in the UK). Steve Brown notes that the changing face of heritage management can be traced through the versions of the *Burra Charter*, away from the idea of the importance of a heritage asset as something 'inherent, immutable or somehow "fixed" within it' [52]. For the *Burra Charter* (and those that draw on it), this importance is termed 'cultural significance'; if, as envisaged, local communities are to have any role in the assessment of that significance, then both the assets themselves and their significance can be expected to change—to be 'mutable' rather than 'mute', to use Brown's terms.

Dirk Spennemann has taken the idea of 'shifting baseline syndrome' from the field of historic ecology and has applied it to heritage studies to explore how the passage of time and intergenerational change impacts the assessment of heritage significance, concluding that significance will wax and wane [53]. It follows that, over time, the mutability of significance may come into tension with the original reasons for inscription on a protected list, exposing unavoidable biases; this is readily visible in the English listing system, the initial phases of which were done at pace from 1944 onwards and which largely covered medieval churches, country houses, and buildings from before 1750 [54].

It seems no accident that many of the authors concerned with change and heritage draw on Australian experience, including of course Laurajane Smith [55], as well as Brown, Spennemann, Harrison, and many others. As Brown points out, the Australian context demands heritage processes that encompass Indigenous as well as more conventional 'European' heritage, generally in the form of historic buildings and sites ([52], p. 21). This reflects the Araoz–Petzet divide discussed in Section 2 above; the inclusion of the UK on the change side of the argument is due to Historic England's adaptation of the Burra framework in *Conservation Principles*, which is explicitly referenced by Petzet in his inclusion of the UK alongside Australia and the USA ([12], p. 54; [13], p. 10).

Furthermore, the generative view of significance also fits well with the narrative approach mentioned above, which stresses the ongoing development of historic buildings [11,56]. On this view, a building's significance is always provisional, because we do not know where its story will go in future—to claim otherwise is to put too great an emphasis on the past and present, on where the story happens to have reached now.

In her 2017 Reith lectures, the late historical novelist Hilary Mantel provocatively suggested that:

> History is not the past—it is the method we have evolved of organising our ignorance of the past. It's the plan of the positions taken, when we stop the dance to note them down. It's what's left in the sieve when the centuries have run through it. It is no more "the past" than a birth certificate is a birth, or a script is a performance, or a map is a journey [57].

Mantel's argument, of course, is against a reification of the past into something fixed and fully defined, echoing Ingold's critique, as discussed above; her fourfold opposition of fixed form (plan/birth certificate/script/map) to lived reality (dance/birth/performance/journey) is as compelling as it is lyrical. I suggest that the same can be claimed, perhaps more strongly still, of attempts to codify the significance of historic buildings that continue to change. A 'generative' view of significance still sees the current state of a historic building—in Mantel's terms 'what's left in the sieve when the centuries have run through it'—as hugely important. It treats the building's history with the utmost respect—after all, how can one understand a story without understanding how it began and developed to this point? However, where the 'subtractive' view treats this as all that matters, the 'generative' view treats any assessment of 'significance to date' as provisional, partial, and incomplete.

It is certainly not being argued that change is a good thing in and of itself—that might have been a core belief of high modernity, but the emptiness of its blind faith in progress is all too evident, and our historic environment bears the scars of such naivety. Rather, the argument is that change should not be ruled out *a priori*—as is the preservationist position, and one seemingly shared by Michael Petzet, as noted above. The question of how historic buildings should be allowed to change is far more complex than a simple question of yes or no. Rather, the task is to decide how a building can be changed without destroying its narrative coherence and integrity, for which a more holistic understanding of significance is required than the 'subtractive'.

Once the legitimacy of change to historic buildings is accepted in principle, the key question becomes what sorts of change should be allowed. Given that each historic building is different, that question will resist any top-down answer. Instead, I have elsewhere proposed that historic buildings (of any age) can more helpfully be considered in terms of an ongoing narrative rather than as a completed artwork [11]. In the case of a church, that is a community narrative written over multiple generations of the core community, and one that awaits further chapters. Once narrative is accepted as the central metaphor, the question of managing change is transformed from attempting to keep change to a minimum to considering how to extend the story well, how to add a worthy new chapter to the existing narrative, when one is called for. In order to do that responsibly, we need to understand the story to date as well as we possibly can, but also to be bold with our current chapter, while having the humility to acknowledge that future generations will wish to add their own chapters.

7. Conclusions

This paper has sought to use the example of historic English Parish Churches and their care and conservation under the Ecclesiastical Exemption to illustrate the shift away from heritage understood as tangible product towards an understanding of heritage as non-tangible process. The Ecclesiastical Exemption has been presented as one expression of a living buildings approach, with the suggestion that a 'subtractive' understanding of significance should be replaced by a 'generative' model. On this latter view, heritage as a whole is understood as an amalgam/hybrid/nexus of tangible and non-tangible—a whole that resists analysis into its parts. The implication for conservation specifically is that the 'living' status of historic buildings is seen as central, not peripheral; to respond positively to this reality requires conservation to reappraise its own narrative, both its genesis and its future direction.

The alternative—defending the preservationist 'conservation paradigm' and avoiding such a reappraisal—threatens significant harm to heritage. By deliberately making change difficult, one of two things can happen: first, living buildings become ossified, ultimately destroying the heritage lying in the joining of tangible and non-tangible; or, second, there is a flight from the material through an understanding of intangible heritage that denies our inherent physicality and the enduring importance of tangible heritage. Conservation is in urgent need of thorough engagement with the non-tangible aspects of heritage; the first step will be to engage with the living nature of tangible heritage.

Funding: This research received no external funding.

Institutional Review Board Statement: Not applicable.

Informed Consent Statement: Not applicable.

Data Availability Statement: Not applicable.

Conflicts of Interest: The author is currently a trustee of the National Churches Trust.

Notes

1   The proposed document was issued for consultation, but never formally published; it would have moved policy away from international principles such as the *Burra Charter*, including demoting communal value to become a subset of historic value.

2   On its reception in England, and the relevance of HUL for conservation as a whole, see [15].

3   The University of York's MA in Conservation of Historic Buildings, now running for 50 years, offers one example of postgraduate study in an archaeology context; for a vision of Conservation without Heritage Studies, see John Earl's misnamed but otherwise excellent [37].

4   It is interesting to note that Malta, for example, combines aspects of British and Italian practice, and has its own version of the Exemption.

5   There is a Searchable Database of Judgments at 'Judgments Index'. Available online: https://www.ecclesiasticallawassociation. org.uk/index.php/judgements/judgments-a-z (accessed on 21 June 2023).

6   The 2021 Update is the source for the figures that follow.

References

1.   Hassan, F. Tangible Heritage in Archaeology. In *Encyclopedia of Global Archaeology*; Smith, C., Ed.; Springer: New York, NY, USA, 2014; pp. 7213–7215. [CrossRef]

2.   Aikawa-Faure, N. From the Proclamation of Masterpieces to the Convention for the Safeguarding of Intangible Cultural Heritage. In *Intangible Heritage*; Smith, L., Akagawa, N., Eds.; Key Issues in Cultural Heritage; Routledge: New York, NY, USA; Routledge: Abingdon, UK, 2009; pp. 13–44.

3.   UNESCO. Convention for the Safeguarding of the Intangible Cultural Heritage. 2003. Available online: https://ich.unesco.org/en/convention (accessed on 2 August 2023).

4.	UNESCO. Yamato Declaration on Integrated Approaches for Safeguarding Tangible and Intangible Cultural Heritage, 2004. In *Proceedings of the International Conference on the Safeguarding of Tangible and Intangible Cultural Heritage: Towards an Integrated Approach*; UNESCO: Paris, France, 2006; pp. 18–21. Available online: https://unesdoc.unesco.org/ark:/48223/pf0000147097?posInSet=3&queryId=39b5fb37-d7d0-4339-8007-dbb9dc308b60 (accessed on 2 August 2023).
5.	UNESCO. Co-Operation and Coordination between UNESCO Conventions Concerning Heritage (WHC-04/7 EXT.COM/9). 2004. Available online: https://unesdoc.unesco.org/ark:/48223/pf0000137634 (accessed on 2 August 2023).
6.	Historic England. *Conservation Principles: Policies and Guidance for the Sustainable Management of the Historic Environment*; English Heritage: London, UK, 2008. Available online: https://historicengland.org.uk/images-books/publications/conservation-principles-sustainable-management-historic-environment/ (accessed on 2 August 2023).
7.	Australia ICOMOS. *The Burra Charter: The Australia ICOMOS Charter for Places of Cultural Significance*; ICOMOS: Burwood, Australia, 2013.
8.	Ministry of Housing. Communities and Local Government. National Planning Policy Framework. 2021. Available online: https://assets.publishing.service.gov.uk/government/uploads/system/uploads/attachment_data/file/1005759/NPPF_July_2021.pdf (accessed on 2 August 2023).
9.	Araoz, G.F. Preserving Heritage Places under a New Paradigm. *J. Cult. Herit. Manag. Sustain. Dev.* **2011**, *56*, 1. [CrossRef]
10.	Lipp, W.; Stulc, J.; Szmygin, B.; Giometti, S. (Eds.) *Conservation Turn—Return to Conservation: Tolerance for Change, Limits of Change*; Edizioni Polistampa: Firenze, Italy, 2012.
11.	Walter, N. *Narrative Theory in Conservation: Change and Living Buildings*; Routledge: Abingdon, UK; Routledge: New York, NY, USA, 2020; pp. 25–27.
12.	Petzet, M. Conservation or Managing Change? In *Conservation Turn—Return to Conservation: Tolerance for Change, Limits of Change*; Edizioni Polistampa: Firenze, Italy, 2012; p. 53.
13.	Petzet, M. *International Principles of Preservation*; Monuments and Sites 20; Hendrik Bäßler Verlag: Berlin, Germany, 2009; p. 9.
14.	UNESCO. *Recommendation on the Historic Urban Landscape*; UNESCO: Paris, France, 2011. Available online: https://whc.unesco.org/document/160163 (accessed on 2 August 2023).
15.	Walter, N. Figure and Ground: An English View of the Conservation of Historic Public Spaces. In *Protection of Cultural Heritage*; Forthcoming. Available online: https://ph.pollub.pl/index.php/odk/ (accessed on 2 August 2023). *in press*.
16.	Polanyi, M. *The Tacit Dimension*; University of Chicago Press: Chicago, IL, USA, 2009; 1966, reprint.
17.	MacIntyre, A.C. *After Virtue: A Study in Moral Theory*, 3rd ed.; Duckworth: London, UK, 1985; Chapter 15.
18.	Hardy, T. Memories of Church Restoration. In *Thomas Hardy's Personal Writings. Prefaces, Literary Opinions, Reminiscences*; Orel, H., Ed.; Macmillan: London, UK; Macmillan: Melbourne, Australia, 1967; pp. 203–218.
19.	Stewart, Z. The Plague, the Parish, and the Perpendicular Style: Theories of Change in Late Medieval English Architecture from John Aubrey to John Harvey. In *Lateness and Modernity in Medieval Architecture*; Sullivan, A.E., Sweeney, K.G., Eds.; Brill: Boston, MA, USA, 2023; pp. 131–152. [CrossRef]
20.	Gadamer, H.-G. *Truth and Method*, 2nd ed.; Sheed and Ward: London, UK, 1989; p. 156, 1960 reprint.
21.	Brand, S. *How Buildings Learn: What Happens after They're Built*; Viking: London, UK; Viking: New York, NY, USA, 1994.
22.	Brandi, C. Theory of Restoration. In *Theory of Restoration*; Basile, G., Ed. and Translator; Cynthia Rockwell; Nardini Editore: Firenze, Italy, 2005; pp. 43–170, 1963, reprint.
23.	Salvador Muñoz Viñas. *Contemporary Theory of Conservation*; Butterworth Heinemann: Oxford, UK, 2005.
24.	Artling Exclusive: Ai Weiwei's Dropping a Han Dynasty Urn', The Artling. 17 February 2022. Available online: https://theartling.com/en/artzine/artling-exclusive-ai-weiweis-dropping-han-dynasty-urn/ (accessed on 2 August 2023).
25.	Ai Weiwei|Han Jar Overpainted with Coca-Cola Logo|China', The Metropolitan Museum of Art. Available online: https://www.metmuseum.org/art/collection/search/78215 (accessed on 2 August 2023).
26.	UNESCO. Convention Concerning the Protection of the World Cultural and Natural Heritage. 1972. Available online: http://whc.unesco.org/archive/convention-en.pdf (accessed on 2 August 2023).
27.	Morris, W. The Society for the Protection of Ancient Buildings Manifesto. 2018. Available online: https://www.spab.org.uk/about-us/spab-manifesto (accessed on 2 August 2023).
28.	Latour, B. *We Have Never Been Modern*; Porter, C., Ed. and Translator; Harvard University Press: Cambridge, MA, USA, 1993; p. 69.
29.	Holtorf, C. Averting Loss Aversion in Cultural Heritage. *Int. J. Herit. Stud.* **2015**, *4*, 405. [CrossRef]
30.	Holtorf, C.J. Is the Past a Non-Renewable Resource? In *Destruction and Conservation of Cultural Property*; Layton, R., Stone, P.G., Thomas, J., Eds.; One World Archaeology 41; Routledge: London, UK, 2001; pp. 286–297.
31.	Holtorf, C. Can Less Be More? Heritage in the Age of Terrorism. *Public Archaeol.* **2006**, *5*, 101–109. [CrossRef]
32.	Harrison, R. *Heritage: Critical Approaches*; Routledge: New York, NY, USA; Routledge: Abingdon, UK, 2013; p. 18.
33.	Fairclough, G.J. Archaeologists and the European Landscape Convention. In *Europe's Cultural Landscape: Archaeologists and the Management of Change*; Fairclough, G.J., Rippon, S., Eds.; Europae Archaeologiae Consilium: Brussels, Belgium, 2002; p. 35.
34.	UNESCO. Vienna Memorandum on World Heritage and Contemporary Architecture—Managing the Historic Urban Landscape. UNESCO World Heritage Centre. 2005. Available online: https://whc.unesco.org/en/documents/5965/ (accessed on 2 August 2023).

35. Bandarin, F.; van Oers, R. *The Historic Urban Landscape: Managing Heritage in an Urban Century*; Wiley Blackwell: Chichester, UK, 2012.
36. Taylor, K. Connecting Concepts of Cultural Landscape and Historic Urban Landscape: The Politics of Similarity. *Built Herit.* **2018**, *3*, 53–67. [CrossRef]
37. Earl, J. *Building Conservation Philosophy*, 3rd ed.; Donhead: Shaftesbury, UK, 2003.
38. Department for Digital, Culture, Media and Sport. The Operation of the Ecclesiastical Exemption and Related Planning Matters for Places of Worship in England. 2010. Available online: https://www.gov.uk/government/uploads/system/uploads/attachment_data/file/77372/OPSEEguidance.pdf (accessed on 2 August 2023).
39. Court of Arches. Re St Alkmund Duffield [2012] Fam 158. 2012. Available online: www.ecclesiasticallawassociation.org.uk/judgments/reordering/duffieldstalkmund2012appeal.pdf (accessed on 2 August 2023).
40. Walter, N. Case Study: St Alkmund, Duffield, and the Ecclesiastical Exemption. In *Narrative Theory in Conservation: Change and Living Buildings*; Routledge: Abingdon, UK, 2020; pp. 110–123.
41. National Churches Trust. The House of Good—The Value of Churches to the UK Economy. Available online: https://www.houseofgood.nationalchurchestrust.org/ (accessed on 2 August 2023).
42. Locke, W.J. Recommendations of the Madrid Conference. *Archit. J. Being J. R. Inst. Br. Archit.* **1904**, *9*, 343–346.
43. Insall, D.W. *Living Buildings: Architectural Conservation: Philosophy, Principles and Practice*; Images Publishing: Mulgrave, VIC, Australia, 2008; p. 10.
44. Ingold, T. No More Ancient; No More Human: The Future Past of Archaeology and Anthropology. In *Archaeology and Anthropology*; Garrow, D., Yarrow, T., Eds.; Oxbow Books: Oxford, UK; Oxbow Books: Oakville, CT, USA, 2010; p. 162.
45. Ingold, T. *Making: Anthropology, Archaeology, Art and Architecture, Anthropology, Archaeology, Art and Architecture*; Routledge: Abingdon, UK, 2013.
46. ICOMOS. *The Venice Charter*; ICOMOS: Paris, France, 1964; art. 5.
47. Poulios, I. *The Past in the Present: A Living Heritage Approach —Meteora, Greece*; Ubiquity Press: London, UK, 2014.
48. Poulios, I. Moving Beyond a Values-Based Approach to Heritage Conservation. *Conserv. Manag. Archaeol. Sites* **2010**, *12*, 170–185. [CrossRef]
49. Wijesuriya, G. Annexe 1: Living Heritage: A Summary. 2015. Available online: http://www.iccrom.org/wp-content/uploads/PCA_Annexe-1.pdf (accessed on 2 August 2023).
50. Chesterton, G.K. *Orthodoxy*; Dodd, Mead & Co.: New York, NY, USA, 1908.
51. ICOMOS. *Guidelines for Education and Training in the Conservation of Monuments, Ensembles and Sites*; ICOMOS: Paris, France, 1993. Available online: https://www.icomos.org/en/charters-and-texts/179-articles-en-francais/ressources/charters-and-standards/187-guidelines-for-education-and-training-in-the-conservation-of-monuments-ensembles-and-sites (accessed on 2 August 2023).
52. Brown, S. Mute Or Mutable? Archaeological Significance, Research and Cultural Heritage Management in Australia. *Aust. Archaeol.* **2008**, *67*, 21. [CrossRef]
53. Spennemann, D.H.R. The Shifting Baseline Syndrome and Generational Amnesia in Heritage Studies. *Heritage* **2022**, *5*, 2007–2027. [CrossRef]
54. About The List | Historic England. Available online: https://historicengland.org.uk/listing/the-list/about-the-list/ (accessed on 2 August 2023).
55. Smith, L. *Uses of Heritage*; Routledge: Abingdon, UK, 2006.
56. Walter, N. The Narrative Approach to Living Heritage. *Prot. Cult. Herit.* **2020**, *10*, 126–138. [CrossRef]
57. Mantel, H. Reith Lectures 2017: Lecture 1: The Day Is for Living. Available online: http://downloads.bbc.co.uk/radio4/reith2017/reith_2017_hilary_mantel_lecture%201.pdf (accessed on 2 August 2023).

Disclaimer/Publisher's Note: The statements, opinions and data contained in all publications are solely those of the individual author(s) and contributor(s) and not of MDPI and/or the editor(s). MDPI and/or the editor(s) disclaim responsibility for any injury to people or property resulting from any ideas, methods, instructions or products referred to in the content.

Article

The Protection of the Historic City: The Case of the Surroundings of the Lonja de la Seda in Valencia (Spain), UNESCO World Heritage

Camilla Mileto and Fernando Vegas López-Manzanares *

Centro de Investigación en Arquitectura, Patrimonio y Gestión para el Desarrollo Sostenible (PEGASO), Universitat Politècnica de València, 46022 Valencia, Spain; cami2@cpa.upv.es
* Correspondence: fvegas@cpa.upv.es

Abstract: In geographical terms, historic cities possess an inertia in regard to the modification of urban function. This explains why buildings may change over time, but the location of the functions remains. For over a thousand years, the city of Valencia has concentrated the commercial activity of its historic centre around the building of the Lonja de la Seda, its surrounding buildings, and its adjacent spaces, streets and squares. Recent constructions coexist with centuries-old buildings, witnesses to the transformations of this urban enclave, which has retained its commercial function. Although the Lonja de la Seda was declared World Heritage by UNESCO in 1996, its surroundings, despite being of interest and closely linked to the protected building, were not. This article analyses the history and evolution of the built fabric and urban spaces of this complex, which represents the nerve centre for commerce in the city of Valencia. This text presents research based on studies carried out directly on the buildings in this context by the authors, as well as indirect examinations of documentation from the archives and the existing bibliography. The aim of this study is to showcase how combining material and documentary studies can lead to a broader definition of the tangible and intangible values of cultural heritage. This, in turn, could lead to the comprehensive enhancement of the historic city, where historic residential fabric and notable buildings are merely manifestations of the process for the construction of the city.

Keywords: historic city; city centre; protection; Valencia; historic dwellings; UNESCO World Heritage; buffer zone; typology

Citation: Mileto, C.; Vegas López-Manzanares, F. The Protection of the Historic City: The Case of the Surroundings of the Lonja de la Seda in Valencia (Spain), UNESCO World Heritage. *Architecture* **2023**, *3*, 596–626. https://doi.org/10.3390/architecture3040033

Academic Editor: Johnathan Djabarouti

Received: 30 June 2023
Revised: 24 September 2023
Accepted: 25 September 2023
Published: 7 October 2023

Copyright: © 2023 by the authors. Licensee MDPI, Basel, Switzerland. This article is an open access article distributed under the terms and conditions of the Creative Commons Attribution (CC BY) license (https://creativecommons.org/licenses/by/4.0/).

1. Introduction, Theoretical Context, Scope and Methodology

The Lonja de la Seda (originally, the Silk Exchange) in Valencia, built between 1482 and 1548, was declared World Heritage by UNESCO in 1996 under the criteria "i: to represent a masterpiece of human creative genius" and "iv: to be an outstanding example of a type of building, architectural or technological ensemble or landscape which illustrates (a) significant stage(s) in human history". These values highlighted by UNESCO in the declaration are clearly evident. However, would this highly spectacular building exist without its surroundings? Or rather, would the Lonja de la Seda make sense without its surroundings? (Figure 1).

1.1. Some Definitions for Further Exploration

The concept of surroundings has progressively been defined throughout the theoretical reflection of architectural heritage. Since the 18th century, interest in history and architecture has gradually encouraged the concept of monument, derived from the Latin term "monumentum" (commemorative monument), which is in turn derived from the verb "monere", "to make you think of something" or "to remind somebody of something". However, over time, this concept linked to a single unique building such as the Lonja, has progressively expanded to cover more open and inclusive concepts and to define assets of cultural interest.

Figure 1. Three images of the Lonja de la Seda in three different contexts: its current context; the simulation of an environment on the edge of the sea and the simulation of a new urban city environment (source: authors).

The Spanish Law for Historical Heritage states that "within the Spanish Historical Heritage, and for the purpose of granting greater protection and safeguarding, the category of asset of cultural interest takes on special value, and this covered movable and immovable property forming part of the heritage which has greatest need for such protection" (Ley del Patrimonio Histórico Español, Ley 16/1985). The Assets of Cultural Interest (in Spanish Bien de Interés Cultural or BIC) encompass all tangible and intangible, movable and immovable manifestations that are witnesses to the historical and cultural values of a given society, in this case, Spanish society. The interest in surroundings or what is found around the monument first, then the asset of cultural interest, is also developed gradually, especially throughout the 20th century.

The first clear formulation on the protection of surroundings can be found in texts by Camilo Sitte (1843–1903) [1].   These ideas, considered by Gustavo Giovannoni (1873–1947) [2], took shape in the Athens Charter (1931), which recommended "that, in the construction of buildings, the character and external aspect of the cities in which they are to be erected should be respected, especially in the neighbourhood of ancient monuments, where the surroundings should be given special consideration." Therefore, monuments need their surroundings and character. The dismantling of the surrounding layout would constitute a loss for the monument itself.

The Venice Charter (1964), which is still widely referenced and present in the current culture of conservation, stated that the notion of monument "embraces not only the single architectural work but also the urban or rural setting in which is found the evidence of a particular civilization, a significant development or a historic event. This applies not only to great works of art but also to more modest works of the past which have acquired cultural significance with the passing of time".

The Quito Charter (1967), geared specifically towards the urban management of historic areas, focused its attention on rural and natural urban settings as assets of importance to heritage. The charter states that "since the idea of space is inseparable from the concept of monument, the stewardship of the State can and should be extended to the surrounding urban context or natural environment" including the cultural assets it contains while also pointing out that "the need to reconcile the demands of urban growth with the protection of environmental values is today an inflexible standard in the formulation of regulatory plans at both the local and the national levels. In this respect, every regulatory plan must be carried out in such a way as to permit integration into the urban fabric of historic districts and ensembles of environmental interest".

These topics were once again vehemently addressed in the context of legislation in the European Charter of the Architectural Heritage (1975), where the concept of architectural heritage (expanding the interpretation of monument) had already been established, through the statement that "the European architectural heritage consists not only of our most important monuments: it also includes the groups of lesser buildings in our old towns and characteristic villages in their natural or manmade settings. For many years, only major monuments were protected and restored and then without reference to their environment. More recently it was realized that, if the surroundings are impaired, even those monuments can lose much of their character." Subsequently, these topics continued to be addressed in charters such as that of the Granada Convention (1985) and the Washington Charter (1987).

A major step was taken in the field of heritage analysis following the Convention for the Safeguarding of the Intangible Cultural Heritage (2003) where material cultural heritage (architectural, urban, movable, immovable, etc.), is awarded equal importance to intangible cultural heritage, which was defined as "the practices, representations, expressions, knowledge and skills—as well as the instruments, objects, artefacts and cultural spaces associated therewith—that communities, groups and, in some cases, individuals recognize as part of their cultural heritage". This new approach enriches the very concept of surroundings, as defined in the Xi'an Declaration (2005): "beyond the physical and visual aspects, the setting includes interaction with the natural environment; past or present social or spiritual practices, customs, traditional knowledge, use or activities and other forms of intangible cultural heritage aspects that created and form the space as well as the current and dynamic cultural, social and economic context".

All these reflections touch upon the essence of "surroundings", where it should be understood that there is no separation between the asset (or monument) and its surroundings (tangible and intangible environment), as the symbiosis between both does not allow their separation. The "monument" exists in relation to the surrounding tangible setting (urban, architectural, material, etc.), and the intangible one (historic, cultural, social, economic, etc.), which brought it about, and the surroundings or environment exist as the result of a culture (historic, urban, architectural, social, etc.), which has resulted in a series of tangible and intangible manifestations, some more notable than others, but equally valuable to the local culture.

1.2. State of the Art, Objectives and Methodology of the Research

From the late 20th century, and particularly the start of the 21st century, increasing numbers of studies have appeared on the architecture of the historic city of Valencia. These have also included several studies on the building of the Lonja de la Seda, detailing research on its history, architecture, construction and state of conservation (including [3–5]). Furthermore, research has also been published on the Plaza del Mercado, the Mercado Central and different markets in Valencia [6–11], as well as surrounding buildings (including [12,13]).

Publications were also found on urban development and urban planning in the city of Valencia during different periods (including [14–17], their archaeology [18,19], compendiums of views and historic engravings [20,21], and historic documents [22], as well as studies on important buildings from different periods in history [23–25]. However, previously the fabric of historic residential buildings had only been examined in a few select cases [23,26].

Given the limited number of detailed studies on historic residential buildings in the city, in the year 2000, the authors of this text began their research on dwellings in the historic city of Valencia, examining them from a historical, architectural, material, constructive, and conservation perspective. This study, developed over almost 15 years and published almost in its entirety in the two volumes of the book "Centro Histórico de Valencia. Ocho siglos de arquitectura residencial" [27] constituted an advance in its field and was recognised with different research awards. The façades of historic residential buildings within the perimeter of Ciutat Vella, or the historic centre of Valencia, considered an asset of cultural interest, were all studied and catalogued in detail, while the urban and architectural development of the city was also expanded on. Studies were also carried out on fully accessible buildings, as well as constructive materials and techniques (rammed earth walls, brick, walls, timber floors and ceilings, renderings and polychromy, joinery, balconies, building interiors, etc.). In addition, a filtering process was carried out on all the construction permit files from the 18th century and on a selection of 19th century files found in the Municipal Historical Archive of Valencia (AHMV onwards).

This allowed the authors to study both the layout of the buildings over time and any transformations observed over the last three centuries. Furthermore, laboratory studies were used to determine the composition of the materials, while dendrochronology was used to date timber elements, and chronotypology to date walls, floors, ceilings, joinery, ironwork, balconies and doors. This detailed material study and the dating of historic residential buildings were cross-referenced with the protection plans in place at the time in order to identify any shortcomings in the protection of these buildings [28]. Part of this study has been incorporated into the new protection plan for the city, which was passed in 2020 [29] (Figure 2).

(a)   **(b)**

Figure 2. (**a**) Map by the authors, superimposing buildings protected until 2019 (with different shaped black symbols depending on the level of protection) and buildings deserving protection according to the authors' research (with different colours used to denote building characteristics). (**b**) Protection plan passed in 2020 including many of the buildings highlighted in the map drawn up by the authors [29].

As mentioned above, these studies generally focused on the area known as "Ciutat Vella", but no specific neighbourhoods in the historic city centre were studied in detail. This article presents later research that partly used already available data, adding new information to focus on a specific area outlined within the historic city of Valencia as a case study: the surroundings of the Lonja de la Seda. This new research aims to combine different historical, literary, artistic, documentary, material, constructive and urban perspectives in order to highlight the material, social, historical, urban, architectural and ethnological values of the historic centre of Valencia.

The case study chosen is the neighbourhood of the market, or in broader terms, the surroundings of the Lonja de la Seda in Valencia (Spain), declared World Heritage by UNESCO in 1996 (Figure 3). The subject of study in this research is not the Lonja as a monumental building of recognised heritage value, but rather everything that surrounds it. For this, a study was carried out on an urban scale on the evolution of the area, based on historic views and urban plans. The evolution of urban spaces such as squares and streets was analysed, while historical, literary and ethnographic analysis was also carried out on the published bibliography, views and historic photographs.

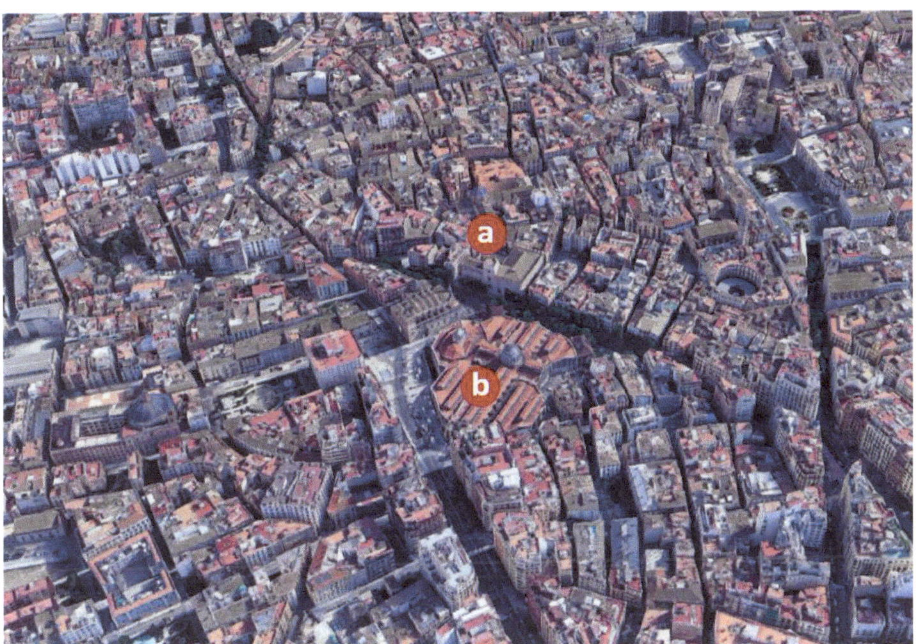

Figure 3. Aerial view of the surroundings of the Lonja de la Seda (**a**) and the Mercado Central (**b**) (Source: Google Earth).

Finally, a material and architectural analysis was carried out on all the monumental and non-residential buildings of the chosen setting, as well as on the residential buildings within the same area. For the purposes of this, analysis is divided into different elements such as squares, streets, emblematic buildings (markets, churches, palaces), dwellings or residential buildings. All these elements, which are analysed from historical, urban, architectural and ethnographic perspectives, provide a wealth of new data as well as a global approach offering a broad-ranging and multifaceted interpretation of the urban surroundings. This research was carried out through a review of the bibliography, the documentary study of archive sources (especially in the AHMV) and the direct study of buildings (architectural types, materials and constructive techniques used, transformations over time, etc.). The combination of several direct and indirect methods in this research has made it possible to identify the tangible and intangible values of a specific case that had not been studied previously—the surroundings of the Lonja—whose historical importance as a market place makes it an emblematic element of the city.

2. The Surroundings of the Lonja: Points for Consideration

In legal terms, the Special Plan for the Ciutat Vella district (passed in February 2020) outlines a perimeter for the surroundings of the Lonja, considered an asset of cultural interest (see fiche C1-13 from the Protection Catalogue; [29]). This perimeter (Figure 4) includes notable buildings such as the Lonja and the adjacent Consulate of the Sea (Figure 4b(a)), Mercado Central or Central Market (Figure 4b(b)), and churches such as those of Santos Juanes (Figure 4b(c)) and Sagrado Corazón de Jesús de la Compañía or Jesuitical church (Figure 4b(d)); public spaces such as the squares of Mercado (Figure 4b(e)), Collado (Figure 4b(f)), Ciudad de Brujas (Figure 4b(g)), of la Compañía (Figure 4b(h)) and some streets and avenues such as María Cristina avenue, and the streets of Taula de Canvis, Cajeros, Danzas, de la Lonja, Pere Compte, Ercilla, Derechos, Vieja de la Paja, Sampedor; as well as the dwellings and buildings included in or adjoining this perimeter.

Figure 4. (**a**) The location of the Lonja de la Seda and its surroundings in the whole of the city of Valencia (red square); (**b**)delimitation of the protection environment of the silk market (identified in red) (source: (**a**) authors; (**b**) [29]).

However, beyond the legal definition of surroundings for assets of cultural interest, and with the freedom afforded by not having to establish regulations, this text aims to analyse certain elements or factors which over time have contributed—and continue to contribute—to the definition of the surroundings of the Lonja. Most of the images of the Lonja produced by 19th century travellers (including Laborde, 1807; Charles de Lalaisse, 1812; Rouargue, 1850) [20,21] show the market square on the right-hand side of a triangular space, where it appears surrounded by residential buildings, opposite the church of Santos Juanes and on occasions the market building. The square is always bustling, full of people and stalls. The view of the 19th century traveller is fixed on the elements that define the surroundings while also providing guidance (Figure 5).

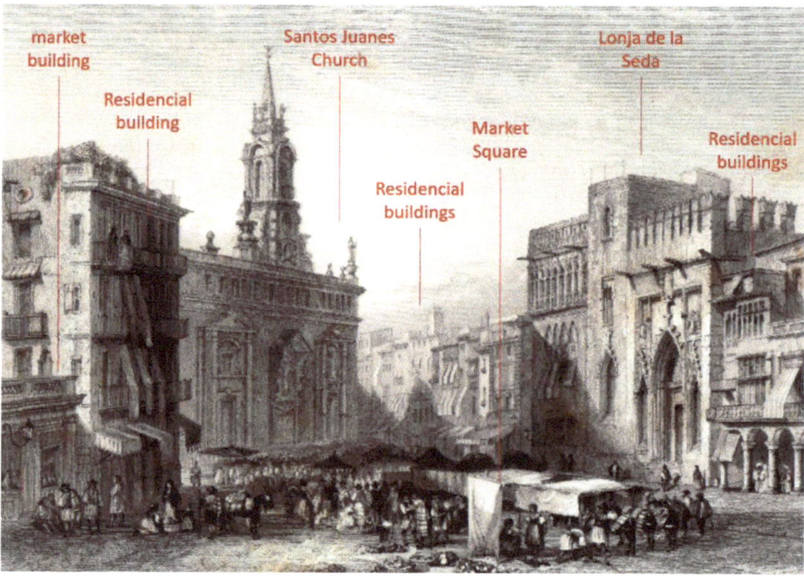

Figure 5. View of Plaza del Mercado (author: Rouargue, 1850 [21]), showing the Lonja de la Seda to the right, the market buildings and the Church of Santos Juanes to the left, and the surrounding residential buildings characteristic of the square's surroundings.

2.1. The Market Square

The Plaza del Mercado or market square in Valencia is an outstanding example of the urban geographical inertia of a function, in this case, the buying and selling of foodstuffs and objects. This is not unlike the inertia of the position of a religious building within a city. As the centuries pass, the container for a function can be replaced or transformed but the activity remains. This is what has happened with the site of Valencia Cathedral, which was previously a mosque, before that a Visigoth basilica, and initially, a Roman temple. The market square has been fulfilling the same function for at least a thousand years.

The market square is without doubt one of the most important elements of the surroundings of the Lonja: a space among buildings that is empty, yet always full of activity and people. Over centuries this square has been shaped and gradually transformed thanks to the new forms of design observed in cities and public spaces, as well as different historical developments and new demands. The market square was defined fundamentally during the Middle Ages, based on the profile of the Moorish town wall on its northeast side. The area occupied by the square has never had a clear and precise geometry. Its outline is the result of the construction of buildings that throughout history have progressively drawn these limits: the churches and convents, the markets, the Lonja and the buildings with shops and dwellings. Therefore, the form and dimension of this square have depended more on its outer limits than on the space itself and, as a result, both its shape and surface have changed over time.

The space currently occupied by the market square was originally outside the walls of the Islamic city of Balansiya (*arrabal* of Boatella), between the Gate of Boatella (*Bab al-Baytala*) and the Gate of the Alcaicería (*Bab al-Qaysariya*). The Gate of Boatella must have been one of the main accesses to Balansiya, used for the passage of goods. Furthermore, near the Gate of the Alcaicería there was market activity, both inside and outside the city walls ([18], p. 425). The market area in the city of Balansiya could therefore be said to have been consolidated after the city's conquest by Jaime I in 1238 and was in the area once occupied by the market of the Islamic city. Outside the wall, opposite the Gate of Boatella, was the *arrabal* of Boatella, a small rural settlement near a large necropolis dating to late antiquity [19]. This *arrabal*, made up of houses inhabited at the time of the conquest of Jaume I and referenced in the *Llibre del Repartiment*, may have been walled, with two accesses, as gleaned from the information from Barceló [15].

Although in the Islamic era, there was already a space outside the walls used as a market, the consolidation and development process began in the Middle Ages, when the space outside the walls was taken over through the construction of convents, the incorporation of the *arrabales* and the foundation of the *pueblas* (medieval urban settlements), which gradually formed the outer perimeter of what later became a medieval city delimited by walls erected throughout the 14th century [30]. This process of progressive colonisation of the territory surrounding the medina led to the definition of the market square through the gradual construction of a number of religious buildings. The foundation in 1240 of the Convent of La Merced ([8], p. 72) was followed by that of the Convent of Santa María Magdalena, and the establishment in 1268 of the former hermitage of Santos Juanes. This space delimited by religious buildings to the southwest and by the Islamic wall to the northeast was gradually incorporated into the medieval city, where increasingly bustling temporary and permanent shops could be found in the triangular area, while artisans and workshops tended to be located on the perimeter. The construction of the new medieval wall and the sizeable expansion of the urban area meant that the market square remained within the geometrical centre of the city.

On the night of 16 March 1447, for a full seven hours, a fire raged through almost two hectares of houses on a corner of the market square. This event was recently studied in depth to document information relevant to the history of this part of the city [31]. The fire, which was probably started as revenge for an execution that had taken place in the same square days earlier, destroyed 46 houses and the fish stalls of the market and resulted in the death of 10 people. According to reports at the time, the fire completely destroyed the carpentry workshop as well as the buildings from the cemetery of Santa Catalina to Puerta Nova, and from El Trenc to the poultry and fish sellers. As Jaume Roig ([16], p. 77) put it at the time: "…La Pellería, Trench-Fusteria, Fins mig mercat, Nas vist cremat."

The number of people receiving damages for the destruction caused by the fire provides a very vivid picture of the range of artisan activities occurring in this district—and so near the market square: carpenters (the guild most affected by the fire), merchants of wool fabric, spices or second-hand clothes, bakers, cobblers, ironmongers, tailors, painters, cheesemakers, box manufacturers, haberdasheries, etc. Furthermore, following the extensive fire damage to the buildings, a plan for urban regularisation was implemented to rectify the existing streets and restructure the area ([31], p. 514). No images of Valencia at this time are conserved, but it is apparent that these works to straighten the streets around the square aimed to improve and consolidate a part of the city that was gaining importance and becoming increasingly representative, a status finally cemented between 1482 and 1548 with the construction of the new Lonja de la Seda.

Valencian humanist Juan Luis Vives (1492–1540), who lived in the city until 1509 and knew the market square, was able to see the completed Contract Hall of the Lonja and the adjacent Consulate of the Sea under construction. Away from his family and city, through his literary alter ego Centelles, he said: "What a large market! What excellent order and distribution of sellers and goods! The scent of these fruits! The variety! What beauty and cleanliness! No orchards can compare to those that supply this city, nor is there diligence to equal that of the market inspection and his ministers so that buyers are not deceived …" ([32], p. 356).

The first existing view of Valencia is that drawn by Anthoine Van Den Wijngaerde (1512/1525–1571) in 1563 [20]. This view shows—albeit unclearly—the market square and the surrounding representative buildings: the religious building on the left, with the lettering "Madalena", is the convent of Santa María Magdalena; in the centre, the caption "Logia" refers to the space at the back of the Lonja de la Seda, possibly as a way to distinguish between the Lonja de la Seda and the Lonja del Aceite (Oil Exchange), which may be the building bearing the same lettering and seen on the left; on the far left, with the caption "S. Joan" we find the church of Santos Juanes; in the centre of the square the author draws a structure that is identified as the gallows ([8], p. 93).

On the plan drawn up by Antonio Mancelli (?–1645) in 1608, in the perimeter of the market square we see the notable buildings of the Lonja de la Seda (Figure 6a(1), the church of Santos Juanes and the convents of La Merced and Santa María Magdalena. Behind the Lonja de la Seda, Mancelli draws attention to the Lonja del Aceite (Figure 6a(2). Mancelli's plan shows the market square as a vast empty space with a gallows in the centre which brings to mind the public executions which took place there. The simple profile of the church of Santos Juanes (Figure 6a(3) is highlighted as it shows the church prior to the remodelling work, which resulted in the new church façade looking onto the market square, as seen in the 1704 plan by Father Tosca. However, the church depicted by Mancelli does not blend in with the square and appears completely disconnected from it, despite helping to define the northwest limit. Nevertheless, the edge of the square appears to be delimited by the notable buildings, especially by residential ones, which are anonymous, and follow the same model of ground floor access and first-floor windows. The author provides a clear simplified representation showing the same pattern of residential buildings that is repeated on every plot: a ground floor door and two first-floor windows.

The following plan of the city of Valencia was drawn by Father Vicente Tosca (1651–1723) in 1704, almost a century after Mancelli's (1608). On Tosca's plan, and in the updated version by José Fortea (1738), it is possible to see the transformations undergone by the market square in the 17th century, a topic widely studied by García Peris ([8], pp. 113–154): the construction of the chapel of Communion of the church of Santos Juanes (1643) (Figure 6b(4); the reconstruction of several collapsed houses in different parts of the square (1663–1666); the addition of the fountain by Juan Bautista Pérez Castiel (1650–1707) in the centre of the square (1672), which remained there until it was replaced by a larger one (Figure 6b(5); the consolidation of the church of La Merced and the remodelling of the cloister (both completed in 1662), as well as the completed construction of the bell tower (finished in 1670) (Figure 6b(6); the demolition of the old church of the convent of Santa María Magdalena (1636) and the construction of the new church (completed in 1679) (Figure 6b(7); the Baroque updates to the church of Santos Juanes and the construction of the façade depicting scenes towards the square (1693–1702), including the construction of the podium which hosted *les covetes* (commercial premises) (Figure 6b(3)—expanded in 1713 to surround the chapel of Communion—thus creating 19 premises for vendors (seen on the 1738 plan). Finally, the appearance of porticoes on the different façades of the blocks that look onto the square should also be noted (Figure 6b(8)) on the buildings adjoining the convent of La Merced, on the two blocks between the convent of Santa María Magdalena and the church of Santos Juanes, as well as on the block found beside the chapel of Communion of the church of Santos Juanes.

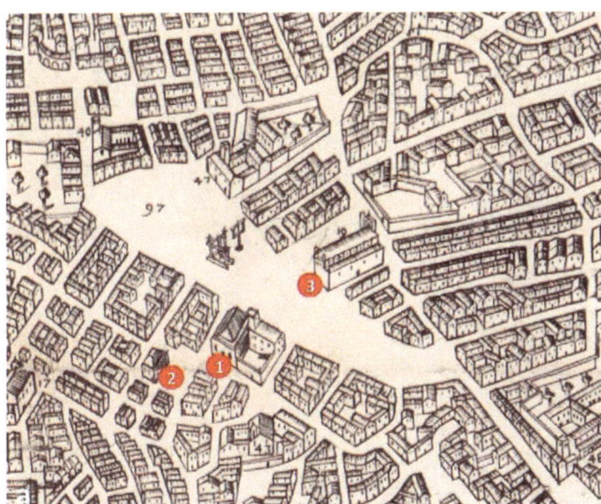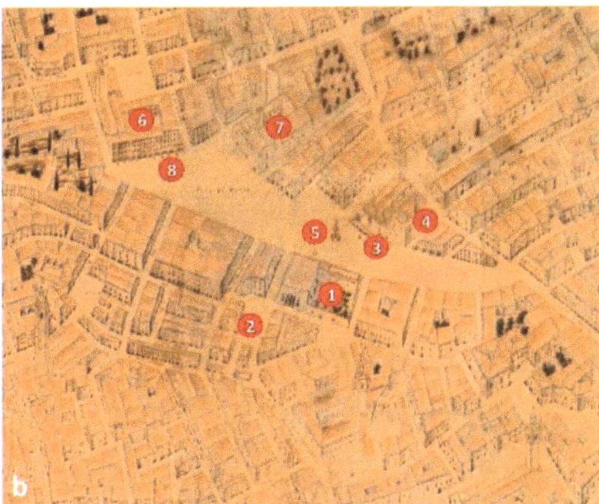

Figure 6. (**a**) Detail of the Market Square and the surroundings of the Lonja on the map of Mancelli (1608) [14]; (**b**) Plano de Tosca (1704) [14]; 1. Lonja de la Seda; 2. Lonja del Aceite; 3. Church of Santos Juanes; 4. Chapel of Communion of the church of Santos Juanes; 5. fountain by Juan Bautista Pérez Castiel; 6. church of La Merced; 7. convent of Santa María Magdalena; 8. porticoes on the different façades.

In addition, the ground floors of the buildings around the square are shown on 18th century plans with continuous doors which never fully become porches but serve as shop windows for businesses. The porticoes of the market square are shown on several occasions throughout the 19th century on plans and in views by different authors. In the "Plano geométrico de la plaza de Valencia y sus contornos" by Francisco Cortés y Chacón 1811 ([14], p. 44–45) the buildings with porticoes on either side in the market square are marked. These can also be seen in the view by Laborde around that time, as well as in later ones such as that by Aulaire in 1830 ([21], pp. 130–131).

After the confiscation of Mendizábal (1836), the market square was modified extensively following the demolitions of the convents on the southern limits of this urban space: the convent of La Merced and the convent of Santa María Magdalena. In 1839, the site formerly occupied by the latter saw the inauguration of a new market called Mercado Nuevo or Mercado de los Pórticos (new or porticoed market). The plot of land that had housed the demolished convent of La Merced was used for the construction of dwellings, which replaced both the convent and the housing with porticoes that had closed off the square on its south border.

These construction projects marked the start of a progressive modification of the urban space which could be described as "from the square to the street" (Figure 7), the empty space by the market buildings was gradually occupied, first by the porticoed market (Figure 7b) and later by the central market, as well as housing. With the construction of the central market (Figure 7c), the market square was reduced to a series of expansions on a single street, taken up for some time by the traffic from carriages, the passing trams and vehicle traffic (Figure 7d). These almost residual triangular spaces completely blurred the concept and use of the former square, which had become unnecessary from the moment the market stalls were relocated to the central market. In this respect, it should be stressed that the project by the architects Peñín and Quintana [33] envisages the pedestrianisation of the market square and a more homogeneous and dignified treatment of this setting, which had gradually deteriorated, and clearly required a solution more in keeping with the heritage importance of the place. However, despite the elimination of pavements in the square, the market square remained a street with expansions, having lost its tangible and intangible dimension as a square over a century earlier.

Figure 7. The evolution of the market square (the colours correspond to the different phases): (**a**) the square and its entrances before the transformations of the 19th century (in green); (**b**) the insertion of the Market of the Porticos (on the plan of 1892, [14]) (in orange); (**c**) the insertion of the Central Market (on the plan of 1892, [14]) (in blue); (**d**) the spaces (in yellow and red) resulting from the transformations of the square and the insertion of the tram (in cian) on the 1944 plan [14].

2.2. Spectacles of Life and Death

The market square, as the main public space in a city where it was once the only square, was the setting for a large number of events, demonstrations and feasts. For example, in 1606, on the first centenary of the canonisation of San Vicente Ferrer, Vicente Colomer designed a mannequin of the saint and made it fly from one end of the market to the other to the amazement of the crowds at what was probably the first zipline in Valencia ([6], p. 144). While the celebration of bull runs, student riots, tournaments and jousting battles was also common, the best spectacle in itself was the daily hustle-and-bustle of the market. In 1494, the traveller Hieronymus Münzer (1495) wrote [34]:

"The inhabitants of the city, both men and women, walk around through the streets at night, and there are always so many people that it looks like a fair is being held (...) I would never have believed that such a spectacle could exist if I had not seen it myself, in the company of my fellow countrymen, the honourable merchants of Ravensburg. The food shops do not close until midnight, so one can buy whatever one wants at any time."

In addition, the function of the gallows, clearly visible in the market square, was to act as a deterrent to prevent inhabitants from carrying out any crimes. Traditional punishments, even for minor crimes, were terrible: lashings; amputation of body parts such as hands, ears, or arms; exile; and even death. Offending the clergy, for instance, was punishable by nailing a hand to wood.

The executioner or *morro de vaques* was a civil servant who was paid a given rate for each of these actions. Although these punishments were applied to all social classes, it was easier for wealthier individuals to evade corporal punishment or the death penalty. According to reports, in 1524 a public gallows with three stone pillars was built so that the provisional timber structure in use there since at least the 14th century, and set up for executions, would not collapse when the hanged prisoner was dropped. In 1599, it was dismantled for the wedding of Felipe III in Valencia Cathedral and was rebuilt shortly afterwards, appearing in Mancelli's plan. In 1632, it was dismantled once again due to King Felipe IV's passing through Valencia ([35], p. 14–19).

The bodies of the dead were sometimes left hanging for hours, or till the next day, to serve as a public example. Although this was to set an example, market stall vendors disapproved as they felt this affected their business negatively. In the 15th century, Jaume Roig ([36], pp. 530–533) stated: *"nor I would eat meat at the market if there were any man was hanging there"*.

The dead prisoners were buried in the sector for hangings in the cemetery of Santos Juanes and later by the ravine of Carraixet ([6], p. 67). In some of the more flagrant cases coming from the Inquisition Tribunal, not only was burial denied, but the bodies were left to be eaten by dogs. This was the case with the last victim of the Inquisition, Gaetà Ripoll, a deist schoolmaster from Ruzafa who was sentenced on 31 July 1826, following a lengthy trial in which he was accused of not going to mass on Sundays, only teaching students the commandments of the Law of God, making them use "Loado sea Dios" ("Praise be to God") as a greeting instead of "Ave María", and not having taken them to adore the viaticum as it passed the school. As a concession to the spirit of the times, he was not burned on the pyre but hanged in the market, over a barrel symbolically painted with flames. After the hanging, he was moved from the gallows at the market to La Pechina from where he was thrown onto the dry riverbed to be eaten by vermin ([17], p. 429).

Prosper Mérimée (1803–1870) made seven trips around Spain between 1830 and 1864 ([37], p. 36). On the first of these, he witnessed an execution in the market square:

"The square was far from full. The fruit and vegetable sellers had not moved from their stalls. It was easy to move around everywhere. The gallows, topped with the Aragon coat of arms, stood opposite the Lonja de la Seda, an elegant building in the Moorish style. The market square is long. The houses which form it are narrow, with several storeys, and each line of openings has a balcony with iron bars. Seen from afar they resemble huge cages. On many of the balconies with bars, there were no spectators. This indifference may be the result, perhaps, of the industrious idiosyncrasy of the Valencian people" ([38], p. 57).

2.3. The Fountain

Whereas a decorative fountain had spouted wine in the market square in 1585 for the visit of Felipe II ([6], p. 159), in 1672 Juan Bautista Pérez Castiel built the first true fountain in the market square, with water driven by a wheel located on Cenia Street (Figure 8a). It was also used to water the garden of the Lonja and supply water to the Guild of Water Carriers or firemen, whose headquarters were in the basement of the adjacent Consulate of the Sea ([6], p. 68). This fountain was replaced in 1852 with a cast iron fountain manufactured in France (Figure 8b). This fountain, which arrived in Valencia by boat, was transported for free from the port to the city centre on the recently inaugurated railway, whose owner, José Campo (1814–1889), had been born at number 80 of the market square ([6], 190) and wished to improve his neighbourhood. In 1875, some sculptures of children and decorative details were added. In 1878, this fountain was moved to Alameda Avenue, beside Mar Bridge, before being moved again in 1933 to its current location, also on Alameda Avenue, beside Aragón Bridge (Figure 8c).

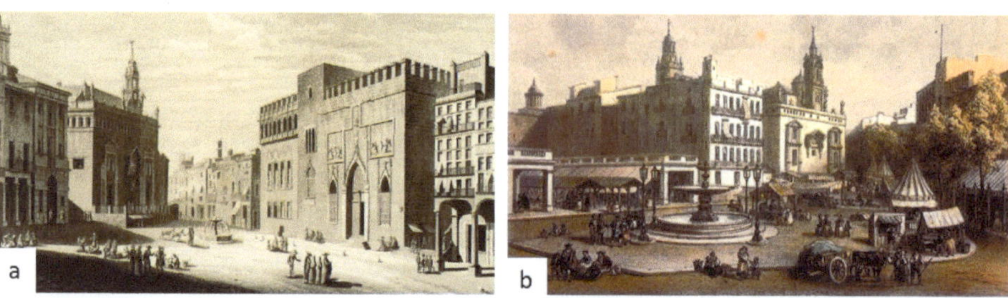

Figure 8. (**a**) The first true fountain in the market square built by Juan Bautista Pérez Castiel in 1672 (drawing A. Laborde, 1811 [21]); (**b**) the cast iron fountain manufactured in France that replaced the first fountain in 1852 [21]; (**c**) the fountain in its current location on Alameda Avenue (source: authors).

2.4. Old Lonja or Lonja del Aceite

This building, which has now disappeared, stood on the current Collado Square and preceded the Lonja de la Seda. It fitted the typology of loggias of the Kingdom of Aragon or Northern Italy, with porticoes on the ground floor and a contract hall on the upper one. The portico was probably built using stone ashlar, while the upper floor was almost certainly brick-supplemented rammed earth. Probably built before 1314, it was successively expanded in 1346 and 1444, and shortly afterwards two atlantes were placed on the corners and became known popularly as Engonari and Engonariesa. Its ground floor archway on three sides had bars added in 1377 and was partially closed off with doors in 1734 to prevent the mess that tended to accumulate inside. On the roof, opposite Derechos Street, stood a stone throne used to display criminals for their public shaming [39].

Once the Lonja de la Seda was built, the Lonja del Aceite was known as Old Lonja, as opposed to the New Lonja. It was later called the Lonja del Aceite or Llotja de l'Oli, although it did not only sell oil but also honey, flour and almonds. The building was so old that it was even referenced in popular sayings: "It is even older that the Lonja del

Aceite" ([6], pp. 39–40). It also hosted the market inspection, which until 1372 was located at Trench Street. Later, it was based in the Longeta del Mustasaf in Santa Catalina Square until 1594, before being moved again, to the Lonja del Aceite, where it remained until 1838. It was then situated in the market square, beside the former porticoed market, until 1916. The Lonja del Aceite was demolished unexpectedly in 1877 and today, only an olive tree planted in Collado Square evokes its presence and disappearance.

2.5. Church of Santos Juanes

This church was founded, following the reconquest of Jaime I, on the site of a mosque outside the city's Moorish walls, yet another example of urban geographical inertia, like the market. This primitive medieval construction was rebuilt and remodelled on several occasions following the fires of 1311 and 1592, the construction of the Chapel of Communion in the mid-17th century and the transformation of the interior vault and construction of a new façade looking onto the market square in the late 17th century [25,30]. Furthermore, in 1700 the church of Santos Juanes applied to the Council of the City of Valencia for the cession of land behind the church, planning to build a terrace to cover the market stalls that had occupied this space since the time of Jaime I. Once the permit was granted on 1st August 1700 ([12], pp. 11–12) the sculptor Leonardo Julio Capuz Calvet (1660–1731) was put in charge of its construction in exchange for 67 annual rents as payment for the construction of the market stalls, their doors and the upper terrace. Eighteen years later the sculptor sold this usufruct to the parish of Santos Juanes ([40], p. 60). During the work, numerous burial sites belonging to the Moorish cemetery of the former mosque or *almacabra* were unearthed. Father Tosca included this terrace with small premises on its lower perimeter in the 1704 plan. These were once known as *les paradetes de Sant Joan* or *les llanterneries* but are currently known as *les covetes de Sant Joan* (Figure 9). The construction of this terrace took up approximately 110 m^2 of public land from the market square.

Figure 9. *Les covetes de Sant Joan*: (**a**) *les covetes* in an 1857 engraving [21]; (**b**) the excavation carried out to bring to light *les covetes* (2021) (source: authors).

It is striking how the increase in the level of the market square over the last 325 years has led to these market stalls becoming semisubterranean. Apparently, this was not the case initially. Recent excavation carried out in the market square for the new paving shows that the bases of the stone ashlar pilasters between the stalls were definitely designed to be seen. Based on the engravings and photographs of the steps to the right of the Lonja, and the descriptions by Llombart [41], it is thought that the level of the square has probably increased by 70 cm since 1700—and probably more going back to 1484—although the increase in height of the paving would have been slower initially and accelerated from the mid-19th century onwards.

This façade of los Santos Juanes and its magnificent terrace, from where the spectacle of life, death, monuments and the space of the market square could be contemplated, suffered extensive damage in the bombing orchestrated by Capitán General Rafael Primo de Rivera y Sobremonte (1813–1902) in the second week of 1869 against the most densely populated areas in the city, as military repression against the federal revolution. Renowned American journalist Henry Morton Stanley (1841–1904), a war correspondent for the New York Herald in Valencia, described it:

"Miraculously, the Gothic Lonja has been spared from damage. However, this has not been the case with the church of Santos Juanes, located opposite. A statue has been knocked over; the niche of the Virgin has been vandalized and defaced; the gargoyles have lost their stone wings, and the church tower has also suffered extensive damage (...) A dozen trees have been cut down in the market. Eight houses there have been rendered unusable; they will have to be rebuilt. The market itself has suffered greatly: collapsed columns, a broken roof, destroyed stalls, etc." ([42], p. 158) (translation by the authors).

In the space of only fifteen minutes, Stanley counted up to one hundred and thirty barricades around the market square. The position of these barricades, put up by Milicia Nacional volunteers and the army of Primo de Rivera, and their evolution in this mini-civil war, which took place between the 8 and 16 October was documented, as can be seen in two plans of Valencia from 1869 ([14], pp. 84–85, 90–91). Stanley also mentioned the existence of graffiti on the Lonja, which at the time was crowned by a flag saying, "Federal Republic", and the wall which closed off the Escalones de la Lonja street: "War on the general and peace to the soldier" and "Death penalty to the thief". The description of the market square following this major battle, combined with the trademark literary style which Stanley imprinted on his chronicles, is chilling:

"What a foul stench! What disgusting stains! What a smell of blood! The blood appears in small puddles, in putrid streams, giving off deadly miasmas. As for the marks of war, it is enough to see the bullet-riddled market stalls. Observe the destroyed balconies, the chipped balustrades, the broken brackets. Look at the walls full of scars. How strange the medieval figures, with their cut-off heads and golden crowns! Even the Virgin has been sacrilegiously and irreparably mutilated, and the Santos Juanes look as if they have been shot for high treason (...) But the scars of the harsh location are far too numerous to list. They are everywhere; they become visible in the ruin and damages of the square, in the fallen trees which once provided shade in the market square, in the horrendous ruins of the surroundings" ([42], pp. 157–158) (translation by the authors).

Twelve days later, on 28 October, Stanley was already in Paris when he received the task of locating and interviewing the missionary David Livingstone (1813–1873). Towards the end of 1871, Stanley found him in Ujiji (Tanzania), where he remained with him until March 1872. A year later, Stanley, a global celebrity at this point, returned to Valencia as a war correspondent to report on the disturbances brought about by the cantonal revolution of Valencia in August 1873, although in his opinion these revolutionaries paled in comparison to those of the federal revolution of 1869.

From its magnificent 300-year-old vantage point, the church of Santos Juanes witnessed and, as seen, was even a victim of, historical developments. Unfortunately, the fact that this vantage point and possible access to the church are now unusable—partly due to the transformation from an urban historic square to a street with expansions—is sadly a wasted opportunity, preventing it from becoming part of the setting of the market square, as well as of the comings and goings of residents and tourists.

2.6. Panses Square, Today Called Compañía Square

This building, located behind the Lonja, was also witness to major episodes in the city's history. Built between 1595 and 1631 on a site specifically selected by San Francisco de

Borja (1510–1572) for his foundation, it appeared to still be under construction on Antonio Mancelli's (¿?–1645) plan from 1608. The Jesuitical professed house or adjoining building in Cenia Street was built between 1668 and 1669 and, as a result, Father Tosca's (1651–1723) plan faithfully reflects the full layout. In 1767, following the expulsion of the Jesuits from Spain on the orders of Carlos III (1716–1788), it seems the church was left half-abandoned and the Casa Profesa was used as the Archive for the Kingdom between 1810 and 1963 [13].

The name of this square, known traditionally as Panses Square, as it was where most of the raisins were sold, eventually changed to Compañía Square. It was customary to read the daily press there. A plaque on the back façade of the Lonja commemorates the 23 May 1808, when the people of Valencia declared war on Napoleon, supporting the command of the *palleter* Vicente Doménech, spurred by indignation at the news published in *La Gaceta* on the abdication of King Fernando VII and the advances of French troops in Spain. The profession of *palleter* consisted in selling sulphur wicks, lighting them to disinfect the empty wine barrels and prevent the proliferation of microorganisms which might alter the wine.

Sixty years later, when the Revolution of September 1868—known as "la Gloriosa"—led to Queen Isabel II (1830–1904) being deposed and exiled, the Junta Superior Revolucionaria de Valencia, headed by the former mayor José Peris y Valero (1821–1877), ordered the demolition of the Jesuitical church [43]. Years later, in 1885, the architect Joaquín María Belda Ibáñez (1839–1912) built a new church, still seen today, following the outlines of the demolished church. The residential buildings on the corners of Lonja Street and Cenia Street and Cordellats Street and Danzas Street, examples to be used for in-depth analysis, have witnessed—as is the case with other buildings—part or all of what has happened in the history of the square or the Lonja itself.

The building at Compañía Square n. 3 on the corner of Cordellats Street is of great interest (Figure 10). This three-storey building is made up of a ground, first and second floor. At first glance, the dressings of the windows on the first and second floors and the railings of the balconies appear to date from the early 20th century. However, this is merely the appearance. In fact, the railing of the first-floor balcony, with its Modernist decoration and lower frieze with bees representing hard work, is made of cast mould, and the solid base of the balcony probably dates from the same period, despite the possibility of a more recent intervention. The balcony windows on the first floor, looking onto Danzas Street, still conserve the valances of die-cut wood board used to protect the external solar protection blinds of the building, another characteristic feature of the Valencia of the time. The wooden joinery, with an upper clerestory and interior shutters, may also date from the same period. In fact, the iridescence of some of the glazing of the clerestories reveals that they were early 20th century blown glass.

However, the upper balcony with rounded corners is much older (Figure 10b), as can be seen from the ceramic tile on the underside of the balcony on Danzas Street, made up of pieces measuring half a palm (11.25 cm), a measure for ceramic production which disappeared around 1740 [44]. A more detailed examination reveals that the railings are wrought iron from a square section and that the lower protective band with flowers framed in rhomboids and a central ring with scrolls, also wrought iron, were subsequently added with small rivets, probably at the time when the lower Modernist balconies were added. It could be assumed that the dressings of the second-floor windows are not Modernist, but date from the 18th century Baroque, although at that point the artisan dwellings were not decorated, far less so on the top floor.

But there are other striking details (Figure 10c): the ground floor in rough ashlar, with a simple corner without decoration and a Roman arch at the entrance. This suggests it was a medieval construction built before the War of the Guilds (1519), after which Roman arches were rarely found at the entrance of residential buildings. The brick construction visible in sections missing rendering is in brick-supplemented rammed earth in dimensions characteristic of the 14th century. In contrast, the industrial brick used to block off the Roman arch is characteristic of the late 19th or early 20th centuries, which is in line with the dates of the interventions on the first-floor balconies. The somewhat irregular distribution of the openings on the ground floor, the windows on the first floor, and the loggia on the second floor with no proposed vertical symmetry, also suggests a building prior to the advent of the Academia de Bellas Artes de San Carlos, which sought geometric order on façades whenever possible. In addition, as will be detailed later on, the 18th century saw the start of a process of completing medieval crown loggias, which were transformed as yet another floor on the building in order to accommodate the city's growing population, a process not observed on this building.

Figure 10. (**a**) The building at Compañía Square n. 3 on the corner of Cordellats Street; (**b**) detail of the balcony on the second floor; (**c**) detail of the ground floor. (Source: authors)

This building is known to have been the residence of doctor and writer Jaume Roig (1407 or 1409–1478). Could this be the same building where the famous author of *L'Espill* lived? It is, in fact, highly probable that at least the walls of the building are mostly those of the dwelling of Jaume Roig. The writer's façade would have been made up of half-body or even full-length windows with wooden railings flush with the wall, with no projections. Undoubtedly, the joinery was made up of blind shutters with no glazing, and in winter wooden frames with oiled linen were used to let in the light while still offering protection from the wind and partly the cold. From the exterior, no evidence is seen of the ceilings of the building, where wooden beams and joists from this period and some details may have survived the passing of time.

The building on Lonja Street no. 8, on the corner of Compañía Square and Cenia Street (Figure 11a), is shown in the 1810 engraving depicting the uprising of the Valencian people against Napoleon and already appears with the same five storeys found today. In fact, the stone doorway with rounded edges; the joinery of the main door; the format of the balcony, projecting three palms (67.5 cm) with iron braces that clear to avoid the openings (Figure 11b); the smaller balconies with rounded corners (Figure 11d); the simple bars of the railings, produced by a blacksmith's hammer and not a cast mould (Figure 11c); the full-body closed-off balconies with a square section and curved corners to avoid the use of the spiked joint, reserved only for the upper edge; and the ceramic tile of the underside of the balconies suggests it dates to 1770–1780 (Figure 11b). The wooden eaves that look both onto Lonja Street and Cenia Street are indicative of a construction which is also characteristic of the 18th century or earlier, as in the 19th century pressure from the Academia de Bellas Artes and local regulations led to the elimination of this type of wooden eave and to the construction of finishing cornices. Under the yellow paint, recently applied, the extensive, slightly irregular and bulging gypsum rendering characteristic of the 18th century is still found, clearly visible.

Figure 11. (**a**) The building on Lonja Street no. 8, on the corner of Compañía Square and Cenia Street; (**b**) detail of the balcony on the second floor; (**c**) detail of the first gate; (**d**) details of the balconies of the upper floors. (Source: authors).

The balconies on the façade of Cenia Street, project three palms (67.5 cm), feature bars, a lower filigree band and a solid ledge measuring a palm and a half (33.75 cm) on the underside of the balcony, suggesting that this façade was remodelled circa 1840. The joinery throughout the building is wooden, with shutters at least on the main floor. When the balconies were added the late 18th century joinery would almost certainly have been large blind wooden shutters where, instead of glass, frames with oiled linen would let in the light in and offer protection from the cold. The use of glass in residential buildings in Valencia took off only after 1840.

However, the stone ashlar corners have a smooth edge and no decoration; a change in wall thickness is detected between the first three floors and the last two; there is a larger separation of the mezzanine between the third and fourth floors; and the ashlar bond on the plinth, which was rougher, is also different from that used in the doorway, which was larger, thinner and only partly connected to the rest of the plinth. This suggests that there may have been an original building with ground, first and second floors which, later, in 1770–1780, was remodelled, adding the large ashlar doorway and two upper floors. All the openings were remodelled, introducing new balconies and closed balconies, a type of intervention that was extremely popular in the city at the time. All these details point to the original existence of a medieval building, probably accessed via a Roman arch where the current doorway now stands; the walls, and perhaps the interior ceiling, were swallowed up by the 18th century intervention, giving the building the appearance it still conserves today.

In conclusion, both buildings appear to have witnessed not only the life of one of the greatest literary figures of the Valencian language and the declaration of war from the Valencian people against Napoleon, but also the construction of the Lonja: Jaume Roig's home, with fewer changes in volume, and the building on Lonja Street considerably heightened by two new floors. Both buildings are therefore several centuries old and are as important as the Lonja both in terms of the context provided and as examples of the evolution of the built material culture of housing within the city. Numerous medieval buildings like these are still conserved around the Lonja, masked to varying degrees by the transformations they have suffered through time.

2.7. The Urbanisation and Buildings of the Market in the 19th and 20th Centuries

This chapter examines San Fernando Street (Figure 12) and the porticoed market, the result of the confiscation and demolition of the two convents, and the central market, which replaced and expanded the porticoed market.

Figure 12. (**a,b**) Buildings with shops on their respective ground floors at San Fernando Street; (**c,d**) building called "El Siglo Valenciano" nowadays ((**c**) exterior; (**d**) interior)). (Source: authors).

2.7.1. San Fernando Street

San Fernando Street was established between the 1820s and 1830s and became a highly successful commercial gallery during the 19th century. The buildings on either side of the street were designed in a homogeneous neoclassical style (Figure 12a,b), with rustication on the ground floor which even spread to the wooden lintels, and three upper floors, the first two of which had continuous balconies projecting three palms with simple bars and struts, as well as dressed balcony windows.

This uniform urban design project was similar to another one dating from the late 18th century, aiming to transform Miguelete Street into a corridor of Neoclassical buildings by the architect Cristóbal Sales (1763–1833), although it was never completed. It is also similar to the projects completed several years later in the nearby Redonda Square (1839) by the architect Salvador Escrig Melchor on the former slaughterhouse; and on Moro Zeit Street, Rey Don Jaime Street and La Conquista Street (1843) on the plot resulting from the expropriation and demolition of the convent of La Puridad by Antonino Sancho Arango (1805–1876) ([27], pp. 54–55, 68, 766, 774).

In just a few decades, this uniform urban layout was destroyed by the invasive design of shop windows, and worse still, the replacement of some of these buildings with different constructions, some very tall. It is currently hard to imagine the urban uniformity, which can only be sensed in some of the façades conserved from that period but interrupted by later buildings.

A notable example among the buildings on this street was "El Siglo Valenciano" (Figure 12c,d), a shop founded in 1879. This was a pioneering textile department store in Valencia, run by Bernardo Gómez. Another unique building from this period was the Hostal del Gamell, lauded in the book on spectacles in Valencia written by Hispanicist Henri Mérimée (1878–1926). He was a first cousin once removed of Prosper Mérimée ([37], p. 39), who, as already said, had also visited the market on his trips to Spain [38].

2.7.2. The New Market or Porticoed Market

A Royal Order of 5 June 1838 awarded the Ayuntamiento de Valencia the former convent of Santa María Magdalena, expropriated three years earlier through the law of Mendizábal. The space created by its demolition and the addition of nearby surroundings provided a plot for the construction of the porticoed market (1839) by the architect Franco Calatayud Guzmán (1795–1854) (Figure 13). Its U-shaped porticoed structure with a trapezoidal annexe, and the porticoes looking onto the market square, earned it the nickname porticoed market. However, as this building could not accommodate all the activity, the market square continued to be filled with awnings or *envelats* covering the stalls that lined the perimeter, as well as the adjoining streets in the *encants* or flea markets, which filled Vieja de la Paja Street and the outer sides of the market. The square, traditionally in the open air with improvised fabric-covered stalls, or at most sheltered under the building porticoes, was partially transformed into a market with a permanent structure and covered stalls. The porticoed market had a surface area of approximately 5000 m^2 and 200 m^2 of this was gained by taking that space from the market square and blocking the opening of Blanes Street onto the square. José Campo, son of the market spice trader Gabriel Campo and future Marqués de Campo, witnessed the birth of this market from the family home, located directly opposite. Once the construction of the porticoed market was completed, work began on the cobbles of the market square ([6], p. 110).

Figure 13. The porticoed market (1839) by the architect Franco Calatayud (Author: Deroy, 1860, [21]).

Vicente Blasco Ibáñez (1867–1928), who was born near the church of Santos Juanes, was a constant witness to the activity and colourful hustle and bustle of this market, as described in his novel *Arroz y Tartana* (1894) [45].

"Beyond, above the muddle of awnings, the zinc roof of the flower market; to the right, the two entrances to the porticoes of the new market, with the low columns painted in bright yellow (. . .). The two servants found their baskets increasingly heavy, and trailed after their mistress through the compact and restless crowd gathering at the entrance to the new market, whose porticoes, in the brightest afternoon sun, were as gloomy and damp as a cave opening (…). It was there that the monotonous lull of the market was most insufferable. The ceiling of the porticoes echoed and amplified the voices of the buyers."

And the fabulous activity of the market on Christmas Eve:

"Good God…! All these people! The whole of Valencia was there. The same thing happened every year on the day of Christmas Eve. That extraordinary market, which remained open until late into the night, became a noisy festivity, the explosion of joy and din of people who among stacks of food and inhaling the scent of the thousands of things which satisfy human voracity, were rejoicing at the thought of the large amount of food to be eaten the following day. In that long square, slightly arched with narrow ends, like a bloated intestine, there were piles of food for days spreading like a rain of nourishment on tables, satisfying the great gluttony of Christmas, a gastronomic feast, which is like the stomach of the year."

At the same time, similar operations for expropriation, demolition and construction were used to establish the Market of La Boquería in Barcelona (1840), which was covered in 1914 by a large iron roof structure, and the Mercado de Abastos or wholesale food market in Cádiz (1838). Both these markets are still in full operation. However, the Ayuntamiento de Valencia soon became aware of the need for a larger—and most importantly, covered—market. The porticoed market continued in operation until it was demolished in 1916 and replaced with the new central market.

2.7.3. Central Market

From 1881, the Ayuntamiento made several attempts to construct a covered central market. In 1883, a competition was called for new designs for the new covered central market, although the winning proposal, by the architect Adolfo Morales de los Ríos (1868–1928) was never completed. In 1907, the architect Joaquín Almarza submitted a design for a prefabricated iron market, although it was rejected as it was considered too small ([23], pp. 185–186).

Finally, a competition for projects was called in 1910, with six participants applying. These included the team made up of Catalan architects Francisco Guardia Vidal (1880–1940) and Alejandro Soler March (1873–1949), both students of the architect Lluís Domènech i Montaner (1849–1923) (who also happened to be Guardia Vidal's father-in-law). Other participants were the pair of Catalan architects Lluís Homs i Moncusí (1868–1956) and Josep Pujol i Brull (1871–1936); Vicente Rodríguez Martín (1875–1933), Valencian architect and one of the main creators of the Valencian Exposition, which was taking place at the time; and Francisco Mora Berenguer (1875–1961), Valencia architect and author of the Palacio Municipal for the 1909 Valencian Exposition.

Francisco Mora probably had a particular interest in this project. In 1907, he had been commissioned to design Casa Ordeig at the market square 13, to replace a former workshop dwelling in the market, the birthplace of Valencian politician Juan Navarro Reverter (1844–1924), Minister for Tax in Madrid and yet another famous person born near the market. The first project by Mora, taking inspiration from the Art Nouveau medievalism of Josep Puig i Cadafalch (1867–1956), was modified by Mora in 1908, reaching its definitive form, a neogothic Art Nouveau similar to that of Lluís Domènech i Montaner [9].

Following a bitter controversy that was not fully resolved, the winning proposal was that of the architects Francisco Guardia Vidal and Alejandro Soler March. Of the rest of the projects presented, only the proposal by Francisco Mora is still conserved. This project clearly shows the plan for the Colón market, designed and built years later in the Ensanche district by Mora, also taking inspiration from the architecture of Domènech i Montaner, but with three naves instead of one. Mora also designed two standalone buildings: a circular flower market reminiscent of Catalonian and Austrian Secession architecture; and a kiosk for selling the typical and local tiger nut milk, with a triangular floor plan with the corners cut off, taking inspiration from similar examples such as Olivella kiosk and Bar Torino at the Valencian Exposition, which was still taking place [10].

The project by Francisco Guardia Vidal and Alejandro Soler March, which was eventually built, was also greatly inspired by the Modernist architecture of their teacher, Lluís Domènech i Montaner. Here, a highly irregular plot is given a polygonal geometry, which produces a striking building, finished off by two brick pavilions to be used as offices. This project for the new market entailed the demolition of two elongated blocks of the market square, adjoining the church of Santos Juanes, and a third smaller one beside what is currently Palafox Street. The construction of the central market, which continued until 1928, encountered several problems along the way, starting in 1919 when the architects who had designed the project resigned from the site management and were replaced by the Valencian architects Enrique Viedma Vidal (1889–1959) and Ángel Romaní Verdeguer (1892–1973). The central market that was finally built (Figure 14) absorbed even more surface area from the market square, which effectively became a street with small expansions. Of its total surface of 8160 m^2 the main body accounts for 6760 m^2 and the fish market for 1400 m^2.

Figure 14. The Central Market today (source: authors).

2.8. The Dwellings

The area around the Lonja, as well as including these major buildings and the public space in the square, is strongly characterised by tall residential buildings with narrow plots, conferring a unique appearance to this part of the city. As mentioned earlier, the surroundings of a building as notable as that of the Lonja are completely inextricable from the monument itself, to the point that its value depends on what is found around it. The buildings of the market with its workshops or displays on the ground floor have been and continue to be one of the most important elements of the setting.

As seen earlier, little attention is paid to detail in these buildings in the first plan of Valencia, drawn by Mancelli in 1608. Mancelli draws all these buildings practically identical, with a gable roof, an entrance door on the ground floor and two windows on the first floor. Although the drawings of the buildings are extremely simple, Mancelli also provides further information: these are narrow plots, with equally narrow façades looking onto the street. The buildings are grouped into blocks where some plots are deeper. As indicated previously, the market area was close to the Moorish wall so the surroundings of the Lonja, as it stands at present, are the result of two different processes. On the one hand, in the area within the city walls, the existing fabric of houses with courtyards and narrow streets is transformed, gradually replacing or transforming the buildings (Figure 15a), while on the other, in the area outside the walls, the territory is colonised by medieval *pueblas* or settlements with parallel homogeneous plots which run perpendicular to irrigation ditches or paths [27] (Figure 15b). In both cases, these "Christian houses" consisted of two floors (ground and first floor), with a space on the ground floor located by the entrance from the street and designed for use as a shop or artisan workshop, a space at the back towards the interior courtyard, used for the kitchen and alcoves, and a room or chamber at the top for use as a bedroom ([17], pp. 86–87) (Figure 15c).

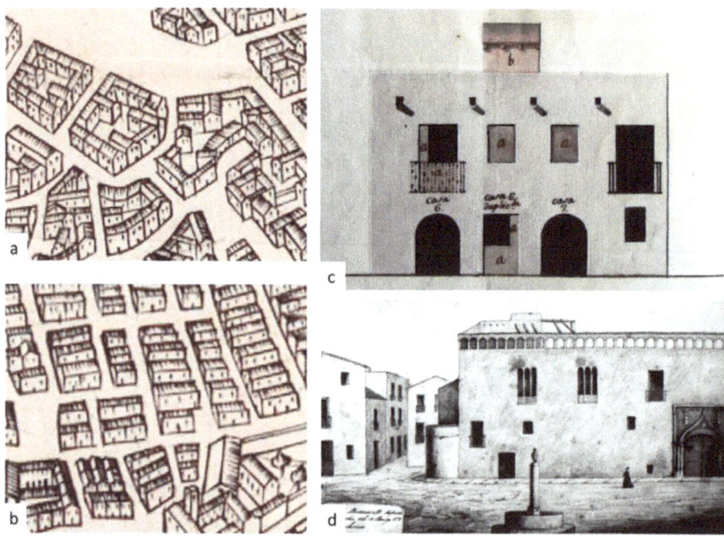

Figure 15. (**a**) Detail of the houses as a result of the transformation of the Islamic city (plan by Mancelli, 1608, [14]); (**b**) detail of the houses resulting from the construction of the new medieval city (plan by Mancelli, 1608, [14]).; (**c**) plan of a house of the time (AHMV).; (**d**) image of the Lonja and the houses that surrounded it [21].

In addition, throughout the 14th century, the Consell worked expressly on straightening and widening streets, creating squares and new streets, closing the cul-de-sacs to prevent waste from accumulating in them [22] or opening them up to create new connections [22]. An interesting example is the order of 1383 to rectify the outline of the streets beyond the Gate de la Boatella ([22], p. 1526). The streets were also organised by eliminating any shutters, benches, protrusions, etc., which invaded them, hindering circulation [22]. As mentioned earlier, in 1447 a fire destroyed some of the houses in this part of the city so the planned reconstruction included blocks with parallel plots open to perpendicular streets.

Therefore, Mancelli's depiction of this part of the city in 1608 is undoubtedly the result of all these operations and although the plan appears limited in detail, the complex blocks stemming from the Moorish structure, the parallel plots resulting from the *pueblas*, and the blocks affected by the fire and reconstructed with blocks between parallel streets can be seen clearly.

On the plan drawn by Father Tosca in 1704, almost a century after Mancelli's plan, the buildings in the surroundings of the Lonja appear greatly transformed. It shows the porticoes on several sides of the market square in buildings with a ground floor and an additional four storeys, constructions that were quite tall for their time. These porticoes (Figure 16c,d) are the ones recorded by the numerous travellers who immortalised the market square in the first half of the 19th century (including: Laborde 1811; Trichon 1834–1835; Chapui 1844) [20]. Evidence of these porticoes is found in the files from the late 18th century in the Municipal Historical Archive of Valencia (Figure 16a,b). In some of these porticoed buildings, applications were made for permits to add balconies while in others the addition of a single storey, changing from a ground floor plus two to a ground floor plus three, was requested [27]. Archive files connected with the market square in the late 18th century include a large number of applications to build balconies or replace old balconies with newer ones (in keeping with the transformations observed throughout the city); demolishing protrusions; reconstructing the exits to the roof and adding railings; demolishing and building new houses with balconies; some inspections for houses at risk of ruin. In 1783, there was an inspection of the portico of a house in the convent of La Merced (AHMV 1783). Different files from 1796, and especially 1797, refer to the repair of damage following the fire which had taken place in the market: "repairing the exit to the roof, repairing windows and other works required following the market fire", "rebuilding a house, remodelling the windows and balconies of the attic, following a fire", "rebuilding a charred apartment and incorporating a balcony", "leaving the house as it was originally", etc.

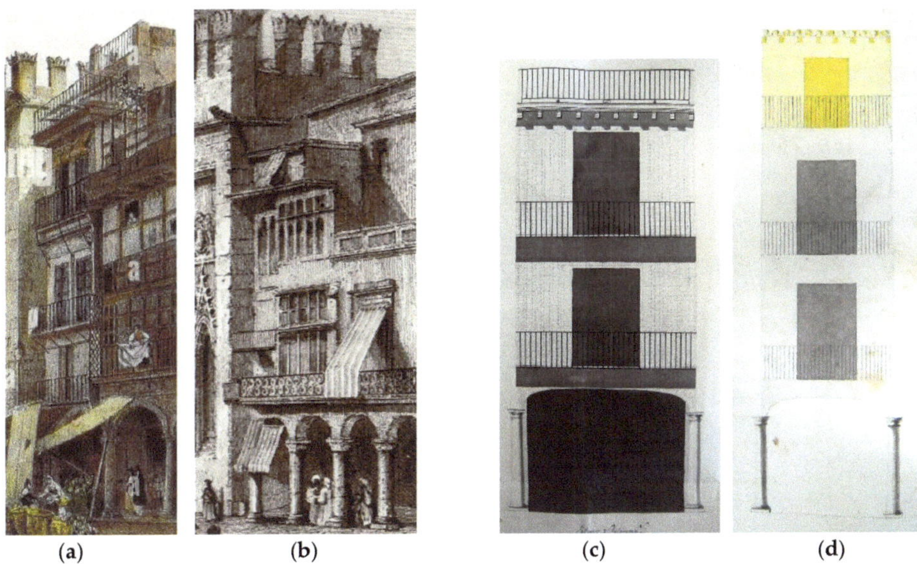

(a)   (b)   (c)   (d)

Figure 16. (**a**,**b**) Two images of the buildings with porticos facing the plaza (from the engravings, respectively, of Chapui, 1844, y Rouargue, 1850, [21]); (**c**,**d**) two drawings from 1797 from the Historic Archive of Valencia that correspond to reform projects for buildings with a portico on the ground floor (AHMV).

In the surrounding streets, Tosca also depicts buildings that are unusually tall for that period, with no porticoes, but with wide openings for workshops or shop windows. All through the 18th century, AHMV files frequently requested permission to work on the doors of the ground floor. There are many files related to the opening of doors, "making a door jamb" or "making two door jambs", in some cases with "ashlar stone", so it is understood that this is about reconstructing one or two door jambs on the ground floor, perhaps to widen them.

There is currently no trace of the porticoes of the market square that disappeared in the second half of the 19th century, as described by Teodoro Llorente in 1889: "the porticoes of the houses disappeared, their narrow windows were widened, the small wooden balconies were replaced with other iron ones, conferring a modern appearance to the entire building". Also, around this time, Blasco Ibáñez described the buildings around the market square: "groups of narrow façades, clustered balconies, walls with lettering, and on all the ground floors, shops selling foodstuffs, clothing, drugs and drink, with doors displaying the names of the establishments, as many saints as are found in heaven and as many common animals of all types are found" [45].

Furthermore, the plans and views from the 19th century show a row of buildings opposite the Lonja and beside the church of Santos Juanes, even seen in photographs from the early 20th century. These buildings were demolished to make space for the construction of the central market. On the corner of this block (identified as number 396 on the plan from 1831) close to the church of Santos Juanes, was the building of the Principal, described by Blasco Ibáñez as "a very poor building, a miserable guardhouse, whose door is guarded by the watchman, weapon in hand, with a bored demeanour, and a bayonet grazing the off-duty soldiers, who are eating their tasteless meal while contemplating the sea of foods which spreads over the square" [45].

In these photographs from the early 20th century, the top end of the same block, looking onto Conejos Street and the porticoed market was made up of a series of narrow houses of different types commonly found in the area of the Lonja (Figure 17): the workshop dwellings, with a workshop and/or shop on the ground floor leading to one- or two-storey dwellings and the chamber or attic store which form part of the medieval legacy of the city (Figure 17a); the stairway houses are small residential buildings, mostly built during the 19th century, combining a workshop or shop on the ground floor with a small entrance to a narrow stair for accessing the independent dwellings on the upper floors, usually one per floor, with minimal dimensions (Figure 17b); the workshop dwelling with a double shop window on the ground floor, possibly the result of merging two workshop dwelling plots (Figure 17c). These can be considered a derivation of the single workshop dwelling given that it functioned in the same way, with access to the upper floors through the shop on the ground floor. In some of the areas surrounding the Lonja it is also possible to find communal residential buildings with two, three or four dwellings per floor with access to the common staircase through a door on the ground floor, usually connected to ground floor commercial premises with shop windows (Figure 17d).

There are currently relatively few examples of medieval workshop dwellings, but some can still be found at different points within the market square and Collado Square and Ercilla Street, Palafox Street, etc. Some of these houses were uninhabited on the upper floors as the living space was accessed through the store. In one of these, a ceiling was identified, with wooden beams, laths and ceramic tile, and given the similarities of its decoration to one of the ceilings of the palace of the Valeriola family, it could date to the 15th century [27]. In some cases, interventions were carried out to establish direct access to the ground floor staircase from the street, separating the living and commercial spaces. Workshop dwellings with two shop windows on the ground floor were extremely common throughout the southeast side of the market square, which featured a series of buildings of this type. Many of these still survive, and both the ground floor shops and the upstairs dwellings have been remodelled.

The surviving buildings, which have undergone numerous interventions and transformations, have become a hybrid mix of architectural styles and elements. Many of these transformations are documented in the files of the AHMV [27] and their history can be briefly examined for a more in-depth understanding of the value of these survivors of history. In the 18th century, closed-off iron balconies were inserted, the loggias or small chamber windows were blocked off, full-length windows were opened onto the balcony, and one or two floors were also added. The widespread presence of balconies in modern Valencia is explained by an 18th century transformation changing a city with plain façades with windows into a city with balconies on all buildings and heights. Closed-off balconies with wrought iron plates and bars, ceramic flooring and struts to support the structure are from this period. In the mid-18th century, the size of ceramic floor tiles changed (from 11.25 cm per side to 22.5 cm per side, or from half to one Valencian palm), lightening the load of the "cages" as fewer plates were needed to secure the tiles. The installation of balconies was associated with opening up windows to create doors for accessing the balconies, in most cases adding joinery, and shutters with no glass, usually folding into the interior side of a thick façade wall seen on the edge of the windows. Interventions like these are widely seen in the buildings around the Lonja, although the examples currently found on the stretch from the market square between the Lonja and Tabla de Cambios Street and between Encolom Street and Bolsería Street are particularly interesting. On these stretches, we find two groups with three and four dwellings which perfectly illustrate these interventions, as well as others from the 19th century (Figure 18).

Figure 17. In the upper part, a photograph from the beginning of the 20th century where the different types of houses in the market area can be recognised; in the lower part, four drawings from the Historic Archive of Valencia that illustrate the different types of houses: (**a**) workshop dwellings (1793); (**b**) stairway houses (1794); (**c**) workshop dwelling with a double shop window (1796); (**d**) communal residential buildings (1854).

In the first half of the 19th century, there continued to be intensive updating of the façades of pre-existing buildings [27]: closed-off balconies were still used, although they were progressively supported by brick brackets which eliminated the need for struts; iron balustrades were added on roofs; in some cases, the access to the living quarters of the workshop dwellings was separate from the ground floor shop thanks to a door added on the façade. The access to the dwelling or dwellings in the upper part of these buildings has always been a weak spot, causing abandonment and deterioration. It is still possible to find abandoned buildings of these characteristics, although possible compatible solutions have been put in place, very similar to those already proposed in the 19th century. In this regard, in Collado Square (Figure 19), the houses on the north side have mostly undergone interventions to divide the commercial openings from the ground floor to create an independent entrance leading to the upper floor dwellings.

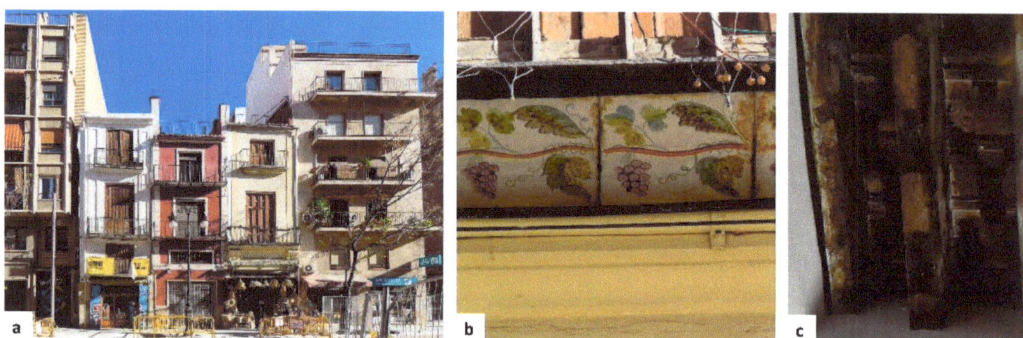

Figure 18. (**a**) Buildings on the stretch from the market square between the Lonja and Tabla de Cambios; (**b**) detail of a balcony from the mid-18th century; (**c**) ceiling with wooden beams, laths and ceramic tile that could be dated to the 15th century. (Source: authors).

Figure 19. Buildings in Collado Square. (Source: authors).

In the second half of the 19th century, but especially in the final twenty-five years, windows with glass were either added outside the shutters or replaced altogether, while wrought iron bars were incorporated into balconies, increasing the potential for decoration and variety. On Palafox Street (Figure 20), it is possible to observe a row of late 19th century workshop dwellings (except for the one on the corner with Blanes Street which is from the late 18th century). These are probably the result of remodelling interventions in previous buildings. In these it is possible to observe how decoration is added to the façade with rustication (used in Valencia from the 1830s), strips separating floors all along the façade, brackets and pilasters adding reliefs to the rendering, which becomes more prominent thanks to colour contrasts; wrought-iron balconies resting on brackets and dating to 1870–1880; and wooden joinery with glazing and *frailero* shutters (which have not been replaced in restoration interventions).

Figure 20. Buildings in Palafox Street. (Source: authors).

Some stairway houses remain on d'En Gall Street, around the corner from Palafox Street (Figure 20). The architectural elements of these two dwellings go back to the second half of the 18th century or early 19th century. Both buildings still conserve the staircase on the façade, visible thanks to the entrance door beside the ground floor shop window and the small windows of the stairwell. Files have been found [27] from the late 19th and early 20th century proposing to move the stairway on the façade closer to the second set of stairs, thus freeing the façade bay and providing a span for a larger room. This type of intervention is usually linked to the widening of the small staircase windows, converting them into windows or full-length balcony doors. Interventions like this can be seen on Pere Compte Street.

Immediately surrounding the Lonja, there are also examples of notable communal residential buildings, probably the result of the merging of medieval plots with narrow workshop dwellings and a remodelling process taking place from the mid-19th century onwards. These buildings include neoclassical mouldings and pediments, wrought iron balconies, and—although covered by paint—the ashlar effect decorating the entire façade as can be seen in photographs from the late 20th century. Buildings of this type, with a communal entrance from the street used to access dwellings, generally two per floor, can be found in several locations around the Lonja, from the market square to Collado Square, and also including Derechos Street or d'En Gall Street, etc.

From the second half of the 19th century, the city of Valencia was immersed in a frenzied rhythm of remodelling and renovation of both the public space and the housing stock. Squares and streets were opened up, new alignments were created in order to widen the small narrow streets of the medieval layout, and older insalubrious residential buildings disappeared to make way for collective housing blocks with improved ventilation and functionality. This process also reached the surroundings of the Lonja. If the "Proyecto de Reforma Interior" by L. Ferreres (1891) had been executed, at present the surroundings of the Lonja would be drastically different and there would be no call for a study of these characteristic dwellings in the market areas. The idea of opening up wide avenues and streets to replace the buildings around the Lonja and the market was a constant during almost half of the 20th century, as recorded in numerous plans and projects. Among these, it is worth noting the plan of "Nuevas líneas para la Reforma Interior de Valencia" by J Goerlich in 1929, which foresaw a complete transformation of the entire area, connecting to Ayuntamiento Square, and opening up the Oeste Avenue behind the central market. The renewal plan, as well as opening Oeste Avenue as far as the church of Santos Juanes, especially affected the southern access to the square from San Fernando Street as María Cristina Avenue and Ayuntamiento Square were now connected. The creation of the new avenue led to the demolition of different blocks of housing up to San Vicente Street and the construction of the collective housing buildings is still seen in the same spot today. These dwellings, mostly built in the 1930s, were on a far greater scale than the construction in the square (although in keeping with porticoed buildings which previously must have been in this area), as well as curved chamfers in stark contrast with the smooth linear façades of this part of the city (Figure 21).

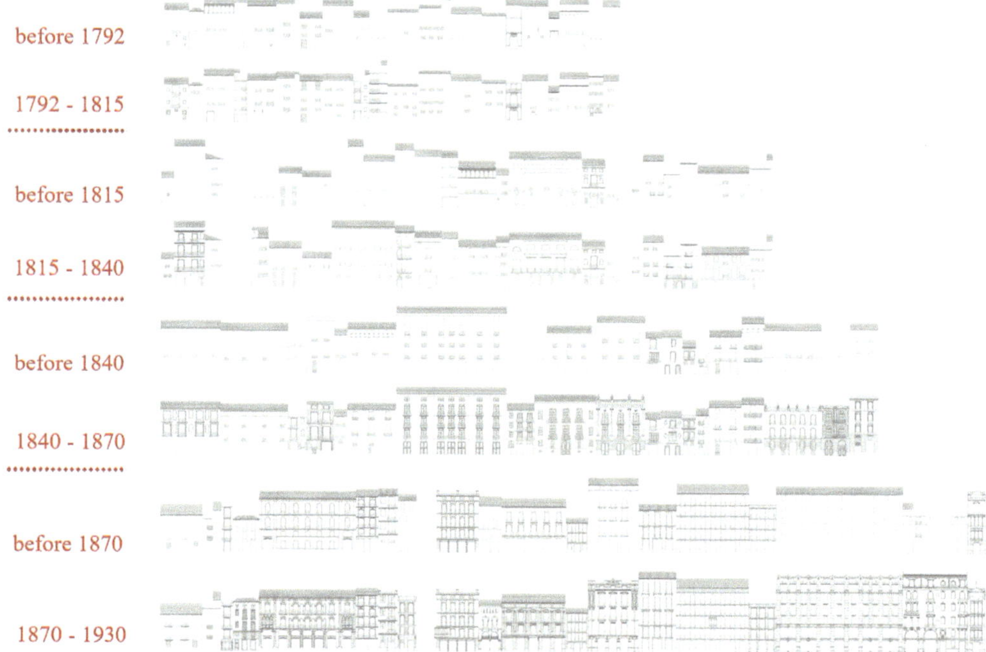

Figure 21. The transformation of the image of the city through the conversion of the facades, demolitions and new buildings during the 1792–1930 period. The reconstruction has been completed through real projects from the files of the Archivo Histórico Municipal de Velencia [27].

3. Conclusions

When the Lonja de la Seda was declared World Heritage by UNESCO in 1996, the building was viewed as an object independent from its surroundings. Between 1996 and 2016, when a review of the documentation for this declaration was carried out, no modifications were made to the perimeter of the declaration, which continued to be linked solely to the building. This article has aimed to prove that the Lonja de la Seda is not an isolated object detached from its surroundings. In addition, its urban setting remains closely connected—in urban, historical, architectural and ethnological terms—to the actual building of the Lonja. The surrounding buildings, whether non-residential (markets and churches) or residential (public spaces such as streets and squares), are all part of an indivisible complex.

The Lonja is part of the enclave of the market of Valencia, which is a thousand years old and has evolved through a history of configuration, transformation and adaptation, which continues even today. This setting is made up not only of monumental buildings like the church of Santos Juanes, the Jesuitical church or the Central Market, but also—and above all—of a residential fabric made up principally of dwellings and ground floor shops woven into it. Furthermore, in most cases, these dwellings, which are several centuries old or even date to the Middle Ages, show a tendency that has favoured the repair, modification and transformation of the building rather than the complete demolition and construction of a new building, so preserving evidence of all the centuries of its history to the present day.

This study has attempted to merge historical, documentary, urban, architectural and material perspectives, employing indirect and direct methods in order to prove that the surroundings of the Lonja are as deserving of attention and protection as the Lonja building itself. The unique architectural and constructive nature that earned the building its designation would not have been possible without its surroundings, linked to commercial activities for years. In this respect, while the Plan de Protección de la ciudad de Valencia (2020) considered these circumstances when establishing an area of protection around the monument (see Figure 4), it is striking that the UNESCO designation for the Lonja does not include a buffer zone, as seen in numerous other cases. Following the research detailed in this text, a proposal is put forward for this perimeter to at least match the protection perimeter established by the Ayuntamiento de Valencia, although this buffer zone could in fact be expanded to cover the entire area associated with the market's activities over the centuries, the artisanal and manufacturing area, and the homes of the merchants and artisans who sold their products from their workshops in the market's area of influence (Figure 22). Finally, it should be noted that the objective set out in this text was the provision of elements for the definition of tangible values (buildings, urban structure, materials and constructive techniques, typologies, etc.), as well as intangible ones (history, transformations, uses, functions, experiences, etc.), for a specific setting within the historic city of Valencia. However, the multi-faceted methodology used in this research could be extended to other settings within the same city, as well as to other historic cities.

The specific case of the Lonja de la Seda of Valencia, declared a World Heritage Site for its exceptional value regardless of its surroundings and the city that has actually produced it, is not an isolated case. Many other outstanding buildings have been declared autonomously and independently of their surroundings and the proposed case aims to stimulate a discussion towards a possible revision of the declarations that does not contemplate the relationship with their respective environments, so that they can be expanded and therefore include the environment not only as a buffer zone but as part of the same declaration. The case of La Lonja shows us how the essence of the city lies in its dynamism. Some things change for the better, others for the worse, and others never change. The work of the guardians of architectural heritage is to ensure that things change for the better, respecting the essence of the city, which is the legacy of generations past and future. Knowledge of the material history of the city through research is essential to earn respect for this built essence in order to protect, restore and enhance it.

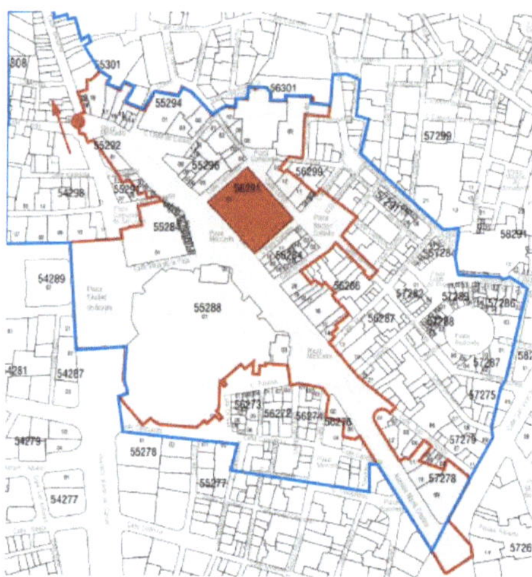

Figure 22. Plan of the surroundings of the Lonja de la Seda with the perimeter of the protection area outlined for the Plan de Protección de Ciutat Vella (2020) (in red) and the authors' proposal for a possible buffer zone (in blue). (Source: authors).

Author Contributions: Conceptualisation: C.M.; methodology, C.M. and F.V.L.-M.; documental research: C.M. and F.V.L.-M.; on site research: C.M. and F.V.L.-M.; photos: C.M. and F.V.L.-M.; writing—original draft preparation: C.M. and F.V.L.-M.; writing—review and editing, C.M. and F.V.L.-M. All authors have read and agreed to the published version of the manuscript.

Funding: This research was self-funded by the authors.

Institutional Review Board Statement: Not applicable.

Informed Consent Statement: Not applicable.

Data Availability Statement: Not applicable.

Conflicts of Interest: The authors declare no conflict of interest.

References

1. Sitte, C. *Construcción de Ciudades Según Principios Artísticos*; Gustavo Gili: Barcelona, Spain, 1980; original edition: 1889.
2. Zucconi, G. *Gustavo Giovannoni. Dal Capitello Allá Città*; Jaca Book: Milan, Italy, 1996.
3. Ramírez Blanco, M.J. *La Lonja de Valencia y su Conjunto Monumental, Origen y Desarrollo Constructivo*; Universitat Politécnica de Valéncia: Valencia, Spain, 1999.
4. Navarro Fajardo, J.C. La lonja de valencia a la luz de las trazas de Montea. In *Arché*; Universitat Politécnica de Valéncia: Valencia, Spain, 2010.
5. Zaragozá Catalá, A. La Lonja de los Mercaderes de València. In *Las Buenas Prácticas en la Gestión del Patrimonio*; Diputación de Alicante: Alicante, Spain, 2021.
6. Corbín Ferrer, J.L. *El Mercado de Valencia. Mil Años de Historia*; Federico Doménech: Valencia, Spain, 1990.
7. Arazo, M.A.; Jarque, F. *Mercado de Valencia*; Ayuntamiento de Valencia: Valencia, Spain, 1984.
8. García Peris, R.C. La Plaza del Mercado de València. Arquitectura, Sociedad e Identidad a Través de Ocho Siglos de Historia. Ph.D. Thesis, Universitat de València, Valencia, Spain, 2020.
9. Vegas López-Manzanares, F. Francisco Mora Berenguer. In *Mercado de Colón. Historia y Rehabilitación*; Ayuntamiento de Valencia: Valencia, Spain, 2004.
10. Vegas López-Manzanares, F. La gestación del Mercado de Colón. In *Mercado de Colón. Historia y Rehabilitación*; Ayuntamiento de Valencia: Valencia, Spain, 2004.
11. Hidalgo Delgado, F. *El Mercado Central de Valencia. Desde su Construcción a su Rehabilitación*; Universitat Politécnica de Valéncia: Valencia, Spain, 2013.
12. Gil Gay, M. *Monografía Histórico-Descriptiva de la Real Parroquia de los Santos Juanes de Valencia*; Auca Llibres Antics/Yara Pérez Jorques: Valencia, Spain, 1909.

13. Medina Ruz, C. *La Fachada Desaparecida de la Iglesia de la Casa Profesa de la Compañía de Jesús de Valencia. Una Propuesta de Restitución Gráfica*; trabajo final de grado inédito; Universitat Politècnica de València: Valencia, Spain, 2020.
14. Herrera, M.J.; Llopis, A.; Martínez, R.; Perdigón, L.; Taberner, F. *Historical Maps of the Town of Valencia. 1704–1910*; Ayuntamiento de Valencia: Valencia, Spain, 1985.
15. Barceló, C. Valencia, espacios y paisajes islámicos urbanos. In *Historia de la Ciudad. Recorrido Histórico por la Arquitectura y el Urbanismo de la Ciudad de Valencia*; Colegio Oficial de Arquitectos de la Comunidad Valenciana: Valencia, Spain, 2000.
16. Orellana, M.A. *Valencia Antigua y Moderna*; Acción Bibliográfica: Valencia, Spain, 1924.
17. Sanchis Guarner, M. *La Ciudad de Valencia. Síntesis de Historia y Geografía Urbana*; IRTA: Valencia, Spain, 2007; original edition: 1972.
18. Ferrandis Montesinos, J. Las murallas de Valencia. Historia, Arquitectura y Arqueología. Análisis y Estado de la Cuestión. Propuesta Para su Puesta en Valor y Divulgación de sus Preexistencias. Ph.D. Thesis, Universitat Politècnica de València, Valencia, Spain, 2016.
19. Martí, J. A la luna de Valencia. Una aproximación arqueológica al espacio periurbano de la ciudad musulmana. In *Historia de la Ciudad II. Territorio, Sociedad y Patrimonio*; Colegio Oficial de Arquitectos de la Comunidad Valenciana: Valencia, Spain, 2002.
20. AAVV. *Les Vistes Valencianes d'Anhonie van den Wijngaerde (1563)*; Conselleria de Cultura, Educació i Ciència: Valencia, Spain, 1988.
21. Catalá, M.A. *Valencia en el Grabado 1499–1899*; Ajuntament de València: Valencia, Spain, 1999.
22. Cárcel, M.M.; Trenchs, J. El Consell de Valencia: Disposiciones urbanísticas (siglo XIV). In *La Ciudad Hispánica Durante Los Siglos XIII al XVI: Actas del Coloquio Celebrado en La Rábida y Sevilla del 14 al 19 de Septiembre de 1981*; Universidad Computense: Madrid, Spain, 1985; Volume 1.
23. Benito Goerlich, D. *La Arquitectura del Eclecticismo en Valencia. Vertientes de la Arquitectura Valenciana Entre 1875 y 1925*; Excmo. Ayuntamiento de Valencia: Valencia, Spain, 1983.
24. Pérez de los Cobos, F. *Palacio y Casas Nobles*; Federico Domenech: Valencia, Spain, 1998.
25. Taberner, F.; Llopis, A.; Alcalde, C.; Merlo, J.L.; Ros Pastor, A. *Guía de Arquitectura de Valencia*; ÍCARO-CTAV: Valencia, Spain, 2007.
26. Simó, T.; Teixidor, M.J. *La Vivienda y la Calle. La Calle Cavallers de Valencia Como Ejemplo de Desarrollo Urbano*; Ediciones Alfonso el Magnánimo: Valencia, Spain, 1996.
27. Mileto, C.; Vegas, F. *Centro Histórico de Valencia. Ocho Siglos de Arquitectura Residencial*; TC Cuadernos: Valencia, Spain, 2015.
28. Mileto, C.; Vegas, F. Blancos en el plano. Edificios desprotegidos del centro histórico de Valencia. In *Actas del Sexto Congreso Nacional de Historia de la Construcción*; Instituto Juan de Herrera: Madrid, Spain, 2009.
29. Esteve, I. *Memoria Justificativa del Catálogo, Pla Especial de Protecció—Ciutat Vella*; Ayuntamiento de Valencia: Valencia, Spain, 2018.
30. Serra, A. Ingeniería y construcción en las murallas de Valencia en el siglo XV. In *Actas del Quinto Congreso Nacional de Historia de la Construcción*; Instituto Juan de Herrera: Valencia, Spain, 2007.
31. Ferragud Domingo, C.; García Marsilla, J.V. The great fire of medieval Valencia (1447). In *Urban History, n.43/4*; Cambridge University Press: Cambridge, UK, 2016.
32. Vives, J.L. *Diálogos: Leges Ludi*; Benito Monfort: Valencia, Spain, 1768.
33. VP/EP. Quintana y Peñín Crean un Porticado Para la Plaza de Brujas y Recuperan "les Covetes". Las Obras, en 2019. Valenciaplaza. 3 June 2018. Available online: https://valenciaplaza.com/quintana-y-penin-crean-un-porticado-para-plaza-de-brujas-y-recuperan-les-covetes-las-obras-en-2019 (accessed on 12 December 2021).
34. Münzer, J. *Viaje por España y Portugal*; Ediciones Polifemo: Madrid, Spain, 2002; original edition: 1495.
35. Solaz i Albert, R. *Valencia Canalla*; Samaruc: Valencia, Spain, 2016.
36. Roig, J. *Spill*; Biblioteca Virtual Joan Lluís Vives: Alicante, Spain, 2001; original edition: 1531.
37. García Cárcel, R.; Serrano Medina, E. *Exilio, Memoria Personal y Memoria Histórica. El Hispanismo Francés de Raíz Española en el s. XX*; Institución Fernando el Católico/Excma, Diputación de Zaragoza: Zaragoza, Spain, 2009.
38. Mérimée, P. *Cartas de España*; Terra Incógnita. José J. de Olañeta Editor: Palma, Spain, 2011; original edition: 1830.
39. Esclapés de Guilló, P. *Resumen Historial de la Fundación i Antigüedad de la Ciudad de Valencia de los Edetanos, Vulgo del Cid, sus Progresos, etc.*; Estévan: Valencia, Spain, 1738.
40. Gayano Lluch, R. *Valencia Retrospectiva*; Biblioteca Valenciana de Divulgación Histórica: Valencia, Spain, 1948.
41. Llombart, C. *Valencia Antigua y Moderna. Guía de Forasteros*; Librería de Pascual Aguilar: Valencia, Spain, 1887.
42. Jiménez Fraile, R. *Stanley. De Madrid a las Fuentes del Nilo*; Grijalbo Mondadori: Barcelona, Spain, 2000.
43. Redacción Eco Republicano. La Junta Revolucionaria de Valencia (1868). In Eco Republicano, 1 March 2021. Available online: https://www.ecorepubliano.es/2021/01/junta-revolucionaria-valencia-la-gloriosa.html?m=1 (accessed on 12 August 2021).
44. Pérez Guillén, I.V. *Cerámica Arquitectónica Valenciana: Los Azulejos de Serie (siglos XVI–XVIII)*; Institut de Promoció Ceràmica de Castelló: Valencia, Spain, 1996.
45. Blasco Ibáñez, V. *Arroz y Tartana*; Biblioteca Virtual Miguel de Cervantes: Alicante, Spain, 1894; original edition: 1894. Available online: https://biblioteca.org.ar/libros/153148.pdf (accessed on 12 August 2021).

Disclaimer/Publisher's Note: The statements, opinions and data contained in all publications are solely those of the individual author(s) and contributor(s) and not of MDPI and/or the editor(s). MDPI and/or the editor(s) disclaim responsibility for any injury to people or property resulting from any ideas, methods, instructions or products referred to in the content.

Article

Spatial Transformation—The Importance of a Bottom-Up Approach in Creating Authentic Public Spaces

Mustapha El Moussaoui 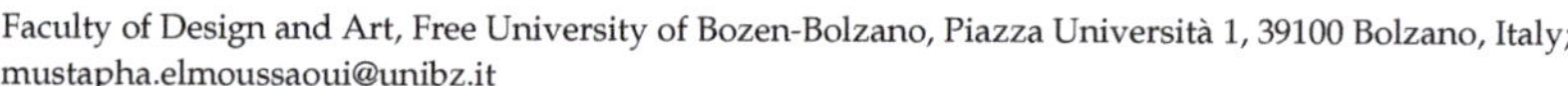

Faculty of Design and Art, Free University of Bozen-Bolzano, Piazza Università 1, 39100 Bolzano, Italy; mustapha.elmoussaoui@unibz.it

Abstract: This study explores the integration of phenomenology in urban placemaking, focusing on the Ghobeiry neighborhood in Beirut. By examining the transformation of a public garden through a phenomenological lens, this research highlights the impact of a bottom-up approach in urban design. The methodology combines a literature review with empirical data gathered from interviews and observations within the community. The findings indicate that the initial top-down development of the public garden failed to resonate with residents, leading to its neglect. However, a shift towards community engagement, initiated by a local social activist, encouraged a sense of ownership and transformed the space into a vibrant, meaningful area. This study contributes to urban planning literature by demonstrating the practical application of phenomenological principles, emphasizing the importance of community involvement in creating authentic urban spaces. It underscores the need for inclusive, participatory approaches in urban development, offering insights into the transformative potential of engaging local narratives and experiences.

Keywords: bottom-up; citizen engagement; phenomenology; city placemaking

Citation: El Moussaoui, M. Spatial Transformation—The Importance of a Bottom-Up Approach in Creating Authentic Public Spaces. *Architecture* **2024**, *4*, 14–23. https://doi.org/10.3390/architecture4010002

Academic Editor: Johnathan Djabarouti

Received: 30 June 2023
Revised: 16 December 2023
Accepted: 19 December 2023
Published: 22 December 2023

Copyright: © 2023 by the author. Licensee MDPI, Basel, Switzerland. This article is an open access article distributed under the terms and conditions of the Creative Commons Attribution (CC BY) license (https:// creativecommons.org/licenses/by/ 4.0/).

1. Introduction

In recent years, the realm of urban planning and design has taken strides towards understanding the nuanced relationship between human beings and their environment [1]. Phenomenology, a branch of philosophy that studies the structures of consciousness as experienced from a first-person point of view, sheds light on the importance of direct, subjective experiences and perceptions in understanding reality [2]. This inquiry lends itself to a study of 'place', a term that transcends physical descriptions and involves the layers of personal and social meanings that individuals and communities assign to a location [3].

The study of 'being in place' from a phenomenological perspective hence takes us beyond the tangible characteristics of a space to consider the lived experiences, emotions, and memories that emerge from human interactions with the place. By doing so, it offers a rich understanding of the place as an interweaving of the physical, social, and psychological dimensions of human existence, potentially providing the basis for more responsive and humane approaches to urban design and planning [4].

One such approach gaining traction is the pursuit of authenticity in placemaking. Authenticity in this context extends beyond originality or truthfulness in a historical or material sense, encompassing a broader understanding that resonates with the 'spirit of the place' or 'genius loci'—a concept that captures the unique, indefinable character and atmosphere that distinguishes one place from another [5,6]. By utilizing a phenomenological appreciation of personal and collective experiences and perceptions, urban designers and planners can attempt to instill or retain the authentic spirit of a place in their projects, by being capable of understanding the socio-cultural aspect of the lived experience of citizens. Moreover, the role of memory is pivotal in this process, intertwining the past, present, and future in a tangible and intangible narrative of space. Memory and reminiscence help to cement the authenticity of a place by preserving its historical narrative and thus perpetuating its unique identity [7]. As individuals and communities interact with their environment, they infuse it with their lived experiences and memories, inextricably binding the human and spatial dimensions. This complex interaction contributes to the formation of a collective memory that is embodied in the 'genius loci', forming an authentic and unique identity for each place.

This study highlights the effectiveness of a phenomenological methodology in comprehending the dynamics of the Ghobeiry–Hay El Jamea neighborhood in Beirut. Initially, the implementation of a public garden in this community followed a top-down strategy, which met with resistance, evident in the frequent vandalism and misuse of the space. However, shifting to a bottom-up approach, where the community members were actively involved in the development process, led to a remarkable transformation. The once resistant community members evolved into caretakers of the public space. This paper explores the transformational impact of community engagement in urban development, using the Ghobeiry case as a focal point. It argues that a bottom-up approach not only raises a sense of ownership among the community members but also facilitates the creation of authentic and sustainable public spaces.

2. Method

The methodology focuses primarily on the intersection of phenomenology and urban placemaking. Initially, the research involves a literature review analysis to define phenomenology within the context of urban studies and to understand the concept and processes of placemaking. This stage aims to establish how placemaking is significantly influenced by the experiences and interactions of individuals within a space. Following the theoretical groundwork, the study transitions into a practical phase with a focused case study of the Ghobeiry neighborhood in Beirut. This component is designed to provide empirical insights that bridge theory with real-life urban dynamics. To achieve this, the research engages with a selected number of residents from Ghobeiry, utilizing phenomenological research methods like in-depth interviews and observations. These interactions are intended to capture the diverse and rich perspectives of the community, offering an understanding of how their everyday lived experiences contribute to the shaping and evolution of a specific public space. Building upon the foundational theoretical and practical aspects of the study, the methodology also incorporates a significant component of direct engagement with key stakeholders and residents of the Ghobeiry neighborhood in Beirut. This engagement was structured to gain a comprehensive understanding of the local urban dynamics from various perspectives. A critical part of this engagement involved conducting structured interviews with members of the local municipality. Over a period of three months, I conducted interviews with five members of the municipality (one of whom is the municipal head, Mr. Maan Khalil), offering insights into the administrative and planning aspects of the region. Additionally, I worked closely on a weekly basis with two members of the municipality. This collaboration provided a perspective on the ongoing efforts and challenges the municipality is facing.

Furthermore, interviews were conducted with 8 families living in proximity to the Hay el Jamea garden of the neighborhood. These interviews encompassed a total of 16 individuals from these families.

In addition to these interviews, I also engaged in informal, open meetings with more than 30 residents from the broader area of Hay el Jamea. These unstructured conversations allowed for a more spontaneous and varied collection of viewpoints.

3. Phenomenology and Being in Place

Phenomenology, as a philosophical approach, underscores the importance of personal perception and experience in understanding reality [8]. In relation to urban environments, phenomenology places emphasis on 'being in place'—a fundamental aspect that molds our relationship with our physical surroundings. As Merleau-Ponty [9] articulates, our body is not merely in space or in time, but it inhabits space and time. Our interaction with urban spaces transcends a mere physical or functional level; we imbue spaces with meaning through our lived experiences, contributing to a multilayered perception of place.

This perspective compels us to recognize urban spaces not just as physical entities, but as sites where personal and collective experiences, memories, and identities intersect. In this framework, the individual and the community become the center of urban planning and design, rather than peripheral considerations. However, understanding 'being in place' is not a straightforward endeavor. It encompasses not just the present experience, but also the complex interplay of history, culture, and personal and collective memory [10]. Here, the work of Bachelard is instructive. His 'poetics of space' suggests that the intimate places of our life—like our home—hold deep-rooted images and memories, shaping our mental constructs of space [11].

Through the lens of phenomenology, it becomes evident that our connection with places is mediated by our senses, emotions, and cognition. Every element of a place, from the materiality of the built environment to the intangible qualities like sounds and smells, informs our perception and experience. In this sense, 'being in place' emerges as an embodied, multisensory experience, rooting us in a specific spatial and temporal context [12]. Heidegger's existential phenomenology introduces a profound dimension to our understanding of 'being in place'. For Heidegger, space is not an abstract entity or mere backdrop against which human life unfolds but is intimately intertwined with our existence. He proposes the concept of 'Dasein', often translated as 'being-there', to underline that human existence is essentially a 'being-in-the-world' [13].

However, Heidegger's 'Dasein' is not an isolated, individual entity, but an involved being, deeply embedded in its world, where the 'world' is a network of meanings and relationships that Dasein comprehends and navigates. In Heidegger's analysis, space is not just an objective, measurable entity, but is inherently relational—we are always in a spatial relation to other beings and things. Thus, our spatiality is a constituent of our being. Moreover, Heidegger emphasizes that our experience of space is shaped not just by physical distances, but by the significance or meaning that entities hold for us. The world of Dasein is not a world of neutral, indifferent objects, but a world of meaningful entities that matter to us, to which we assign importance and value. For instance, a place where we grew up may seem 'closer' to us in a meaningful sense, despite being physically far away. In this endeavor, being in place emerges not just as a physical or sensory experience, but as an existential condition that involves understanding, concern, and care. It suggests that placemaking should be more than creating aesthetically pleasing or functionally efficient spaces; it should aim to create meaningful places that resonate with our lived experiences and existential concerns.

4. Authenticity and the Spirit of Place

Our existential understanding of place sets the foundation for a discussion on authenticity and the spirit of place, or 'genius loci'. Norberg-Schulz introduces 'genius loci' as the particular character or atmosphere of a place, and argues that understanding this character is vital for creating places that are meaningful and authentic [5]. Authenticity in place refers to the qualities that make a space genuine, unique, and meaningful to its inhabitants. It includes aspects such as the history of the place, the cultural and social practices associated with it, and the collective memories and experiences of the people who inhabit it [4]. Authentic places are those that resonate with our experiences and values and evoke a sense of belonging and identification [14]. Creating authentic places, then, is not just about design and aesthetics, but about fostering connections between people and their environment. It involves integrating the physical, cultural, and social elements of a place in ways that reflect and enhance its unique character and history [15]. This is where the principle of 'genius loci' comes into play. Moreover, the spirit of a place is not a fixed or objective entity, but a dynamic and subjective phenomenon that emerges from the interaction between people and their environment [5]. It reflects the unique ways in which a place is perceived, experienced, and valued by its inhabitants. As such, understanding the 'genius loci' of a place requires an empathetic and holistic approach, one that considers not just the physical attributes of the space, but also the meanings, emotions, and memories associated with it.

This emphasis on authenticity and 'genius loci' highlights the importance of place-making approaches that are participatory and people-centered [16]. Such approaches involve engaging with local communities, understanding their needs and aspirations, and integrating their insights into the design and development of public spaces. They aim to create places that are not only functional and aesthetically pleasing, but also meaningful and authentic, places that enhance the well-being and quality of life of their inhabitants.

5. Memory and Authenticity

Building upon our understanding of the phenomenological experience of place and the importance of authenticity in placemaking, we now turn to the role of memory in defining the authenticity of a space. Memory, both individual and collective, plays a crucial role in our relationship with space, shaping our perceptions, experiences, and identities. Tuan posits that "space" becomes "place" when it is imbued with human experience and memory [17]. Similarly, Bachelard in his seminal work, "The Poetics of Space", explores the intimate connections between memory and space, arguing that our most profound, lived experiences are often tied to specific places [11]. These places, imbued with our memories, become repositories of our histories, identities, and emotions, carrying a sense of familiarity, comfort, and belonging. Hence, memory, in this context, is not merely retrospective; it is a dynamic process that shapes our present experiences and future anticipations [18]. In terms of placemaking, memory serves as a critical link between people and their environments, informing the meanings they ascribe to spaces, their emotional attachments to them, and their interactions with them.

The concept of 'lieux de mémoire' or 'sites of memory', introduced by Nora, further illustrates the intertwining of memory and space [19]. Nora suggests that certain sites, such as monuments, landmarks, or even less tangible entities like rituals and symbols, serve as repositories of collective memory, embodying shared histories and cultural identities. These 'sites of memory' are crucial in maintaining a sense of continuity and coherence in the face of rapid social and spatial changes. Therefore, understanding the role of memory in the experience of place offers valuable insights for placemaking. It suggests the need for placemaking approaches that respect and incorporate the historical and cultural layers of a place, preserving its 'memory traces' [20] and promoting a sense of continuity and identity. This might involve preserving historical structures, celebrating local traditions, or creating spaces for community storytelling and commemoration. Moreover, placemaking should also enable the creation of new memories by facilitating social interactions, community activities, and personal experiences. In this way, placemaking can contribute to the ongoing narrative of a place, maintaining its authenticity while allowing it to evolve and adapt to changing circumstances.

6. Spatial Transformation and Placemaking

In the quest for authentic placemaking, the role of spatial transformation and the involved stakeholders cannot be overlooked. Urban theorists and planners have long grappled with questions regarding the dynamics of spatial transformation and the creation of meaningful, vibrant spaces. Central to these debates is the distinction between top-down and bottom-up approaches to urban planning and placemaking.

In a top-down approach, decisions about spatial transformation are typically made by a centralized authority, often with minimal input from the community. While this approach can be efficient and cohesive, it often neglects local nuances and disregards the lived experiences and preferences of community members. As a result, these transformations may fail to resonate with the local population, and thus, might not engender a genuine 'spirit of place'.

In contrast, a bottom-up approach privileges the input and engagement of community members in decisions regarding spatial transformation. Rooted in the belief that those who live, work, and play in a space have the most intimate knowledge and stake in it, this approach emphasizes participatory planning and design processes. As Healey suggests, such a strategy recognizes the pluralistic, multi-voiced nature of the city, respecting the diverse needs, desires, and visions of its inhabitants [21].

French sociologist Henri Lefebvre's (1996) *'Writings on Cities'* underscores the importance of bottom-up processes in spatial transformation. Henri Lefebvre's theoretical perspective on the production of space plays a significant role in understanding urban transformation [22]. Lefebvre contends that space is not a static entity, but rather a socially produced phenomenon, generated by and intertwined with the complexities of social interactions, power dynamics, and economic systems. His triadic model of perceived, conceived, and lived space offers a holistic approach to understanding spatial contexts. This model suggests that space is simultaneously a physical reality (perceived space), a mental construction or representation (conceived space), and imbued with individual and collective experiences and symbolism (lived space). This understanding of space as socially produced posits that changes in society are intricately linked to the transformation of space itself, reinforcing the essential role of individual and collective action in shaping urban environments.

By allowing residents to influence the design and use of their spaces, a bottom-up approach can foster a sense of ownership and attachment, key ingredients in the creation of a true 'genius loci'. Such participatory practices can help ensure that the built environment reflects the collective memory, culture, and identity of its inhabitants, thus enhancing its authenticity and sense of place [23]. However, the effective implementation of a bottom-up approach to spatial transformation and placemaking is not without its challenges. It requires an open, flexible planning system capable of accommodating diverse perspectives and facilitating meaningful public participation. It also necessitates a shift in mindset among planners and decision-makers, from viewing the public as passive recipients of design to active contributors and co-creators of space.

7. Case Study—Hay El Jamea, Ghobeiry

This bottom-up, top-down dichotomy approach was witnessed during our research in Hay El Jamea in Ghobeiry, Beirut. To understand this change we will look a little into its history.

After World War II, Beirut transformed into a bustling center of economic activity, attracting Lebanese citizens from the villages who were seeking better prospects. Unfortunately, the city's economic prosperity was accompanied by visible signs of social inequality, which drove these migrants to seek affordable housing in the suburbs near the city [24]. As Beirut's urban landscape rapidly developed in the 1950s and 1960s, a new era of modernity emerged, accompanied by new urban policies and regulations [25]. Unfortunately, the situation took a turn for the worse, first in 1977, with the start of the Lebanese civil war, then in 1982, with the Israeli invasion of Southern Lebanon, which caused citizens from Southern Lebanon to migrate to Beirut and settle in areas where other communities had already established a presence.

These historical events set the stage for the spatial production of neighborhoods like Hay el Jamea in Beirut's southern suburbs. Despite the intensification of neoliberal practices in the Lebanese economy after the 1990s, and the effects of globalization, these communities continued to maintain their cultural values [26]. It is important to note that the spatial production of Hay el Jamea and other neighborhoods in the region was shaped by various factors, including migration patterns, economic trends, and urban policies. These factors worked together to create a unique urban landscape that reflects the community's values and aspirations. Despite the challenges and difficulties faced by the residents, they managed to create a vibrant and dynamic community that continues to thrive to this day.

8. Space Formation

While doing the research in the neighborhood, to understand how these working-class communities constructed their communal space, and partake in the community events, a phenomenological analysis was performed by interviewing residents who live in the neighborhood for the second degree. The analysis required us to go back and understand their history on an individual level and how most of them came to the neighborhood; moreover, understanding their political and social interactions through in-depth talks.

C05, C08, and C02, are all family members living in Hay el Jamea Neighborhood. These individuals all come from a poor background, are mostly working class, underpaid individuals.

"My father was a gardener for the minister, Rachid Youssef Beidon, in a village in the Bekaa valley known as Janta. Janta served as a trade route between Beirut and Damascus via Yahfoufah's train station. After years of working in Janta, My father requested to move to Beirut to work as a gardener for Beidon's mansion in Beirut during the early 1950s". C05

"My father first came from Sareein (Bekaa Valley), he came to work for Gandour (Lebanese sweets factory). First he lived with his friend with 6 others from Sareein in one room". C08

"I came in the early 1973, first I worked in the port, but I wasn't lucky, as only 2 years afterwards, turbulences started to occur on the way down to work, I had to work as a car mechanic in the neighborhood to avoid the road" C02

However, in the context of civic life, residents demonstrate robust social cohesion, as evidenced by organized neighborhood interactions and governance structures. Local community leaders commonly spearhead the convening of meetings in designated communal spaces within their jurisdiction. These locations can be diverse in their initial intent, ranging from enclosed interior environments to open plots of land, transitional spaces, rooftops, or even adjoining thoroughfares. C12 and C08 mention that in this neighborhood, they meet on weekly basis. The locality features two principal communal venues: one is politically aligned with a local political faction, while the other serves as a versatile space. The latter primarily functions as a forum for residents to discuss communal needs and challenges, partake in collective celebrations, or engage in religious observances such as "Majles azaa" during the Ashura event (a yearly event that takes place for the commemoration of Imam Hussein). When asked if they do some of those meetings in Hay el Jamea garden, C12 explains that not all community meetings are the same; some are more exclusive—which are more of a political meeting—while others are more of a community meeting for needs and problems, and in all these aspects—although this park (Hay el Jamea garden) is now open, still, community members never meet there—they continue to use the old meeting spots. It is possible to assume that this park is not used as a communal meeting space due to its proximity and non-private geographic location. C09, a current community representative, explains that the community preferred using interstitial in-between spaces due to their intimate nature and structure (more private).

In 2014, the Ghobeiry municipality initiated the construction of the Hay El Jamea Garden, aligning with their strategy to augment the region's green spaces. The project commenced with the strategic reclamation of peripheral lands around Hay El Jamea. This process involved a meticulous series of acquisitions, mergers, and reorganizations of these lands, which ultimately led to the transformation of the area into a public park. By the end of the year, the project reached completion, marking the inauguration of the first public park in the neighborhood. Prior to this, local residents had limited exposure to communal recreational spaces. Traditionally, the community engaged in the cultural practices of spending weekends and holidays in their villages of origins (Bekaa valley, or South Lebanon), as noted by C04, C07, C08, and C012. These visits were not just social gatherings but also opportunities for the locals to collect homegrown produce and strengthen familial bonds.

When the municipal authority established the Hay El Jamea Garden (see Figure 1), it was observed that the local residents, residing in the vicinity of the park, persistently utilized the area as a site for garbage disposal and dumping (refer to Figure 2). Furthermore, there was a noticeable trend of deliberate damage to the park's infrastructure. This behavior can be attributed to a lack of a sense of ownership or connection with this newly developed public space, which was imposed in a top-down manner. It is important to note that these residents have a history of independently managing their community spaces, adapting and cultivating areas without external intervention or support from public authorities.

Figure 1. Hay el Jamea's garden after reformation. Source: author (2018).

Figure 2. Hay el Jamea's garden before construction, with the garbage surrounding the old, abandoned structure. Source: Ghobeiry Municipality (2014).

M02, a social activist and member of the municipality council, proposed to create a communal meeting and talk to community members in order to find solutions for this ongoing problem. The residents refused to participate and cooperate at first, particularly because she was a representative of an official governmental body. To penetrate the social border, M02 mentions:

> *"I knocked on every door of each community member and offered coffee and mana'aesh (a traditional Lebanese breakfast pastry made with thyme). Initially, I only wanted to enjoy a cup of coffee with them to get to know them better and create peaceful relations. After two months of these continuous interactions, I created a successful social bond of respect with the community members". M02*

M02 notes a subsequent pivot in her dialogues to focus on the garden initiative. She underscored the uniqueness of this communal asset, further granting residents access via a key to the main entrance, and the power to have it as their own garden, by planting and changing the urban furniture (where it is possible) in the way they would like. Gratifyingly, this sparked a positive shift in the resident's sentiment towards the area. Mobilizing collective efforts, the community transitioned the space from a neglected, graffiti-ridden wasteland to a well-maintained garden.

"I used to toss my cigarette butts here like it was no big deal. Never really thought about it, to be honest. But now, seeing my kids play here every day? Kinda changes your perspective. It's not just some dump anymore; it's where the neighborhood hangs out. Makes you realize what can happen when folks come together to clean things up. I'm glad it changed; it's better for everyone, especially the kids". C18

"I never really paid much attention to this lot before, just passed it on my way to work. It was a real eyesore, if I'm being honest. But now, it's like a little oasis or something. You see families out here, kids playing—it's become a part of our daily lives. It's surprising how a little effort can turn something neglected into something so valuable to the community". C03

"Surely, this land changed a lot, but I still don't use it, Arguile (Shisha) is not allowed in". C07

As the narrative of the Hay El Jamea Garden evolves, it is evident that the space has transformed significantly in the eyes of the local community. Children and families now regularly utilize the garden as a recreational area, a testament to the collective efforts that turned a once neglected space into a vibrant, communal asset. Despite this positive development, it is important to acknowledge that certain aspects of community life remain unchanged. Specifically, the practice of holding community meetings, whether they are politically oriented or centered around communal needs, continues to be conducted in more private settings. These gatherings, steeped in tradition and a sense of discretion, persist in spaces that offer the privacy and familiarity conducive to such discussions.

9. Discussions

Urban environments, as dynamic entities, are shaped not only by their physical constructs but also profoundly by the lived relation between human behavior and the built environment. This research looks into urban placemaking through a phenomenological lens, with a specific focus on how grassroots, bottom-up approaches can infuse authenticity and vitality into public spaces.

One of the findings of this study is the revelation that the initial approach to urban planning, characterized by a top-down directive in establishing public spaces such as the garden, initially failed to resonate with the local community. This disconnect can be attributed to a lack of participatory engagement and a sense of collective ownership. However, a paradigm shift towards a community-driven, bottom-up strategy, spearheaded by a proactive social activist from the municipal council, marked a turning point. This participatory methodology fundamentally changed how community members approached and interacted with the space. The result was the emergence of a public garden that became a meaningful and integral part of the community's daily life.

Expanding beyond its initial scope, this study also highlights the complexities of urban community dynamics. It showcases how historical, socio-cultural, and economic factors intertwine to shape communal perceptions and the utilization of urban spaces. For instance, the garden's transformation from a neglected plot to a working communal space is not just a tale of physical redevelopment but also a narrative of socio-cultural evolution. The community's initial indifference, rooted in a historical context of self-managed communal spaces and a cultural inclination towards private gatherings, gradually gave way to a collective realization of the garden's potential as a communal asset.

The originality of this research lies in its empirical manifestation of phenomenological principles within the realm of urban planning. By bridging the theoretical and practical realms of placemaking, this study provides profound insights into the creation of authentic urban spaces through a bottom-up approach.

10. Conclusions

Cities and urban environments, as we know, are not just physical entities. They are, above all, socio-spatial constructs imbued with meanings and values that shape and are shaped by the individuals and communities that interact with them. The inextricable intertwining of physicality and lived experience in urban landscapes forms the crux of this discourse.

The overarching inference from this discussion is the need to foreground people and their experiences in the process of placemaking. It is through such an approach that urban environments can foster a sense of belonging, facilitate social interaction, and ultimately, serve the diverse needs and aspirations of their inhabitants. Therefore, the pursuit of more authentic and meaningful urban environments warrants an intertwining of phenomenological perspectives with conscious, inclusive, and participatory spatial practices.

This multi-dimensional understanding of placemaking illuminates the path towards the creation of urban environments that are not just physically appealing but also emotionally resonant, promoting a deeper, more fulfilling sense of place for all city dwellers.

The study of Hay El Jamea's transformation provides critical insights into the dynamics of urban development and community engagement. The shift from a top-down to a bottom-up approach in urban planning, as exemplified in this case studied from a phenomenological approach, underscores the profound impact of involving local communities in the shaping of their environments. The transformation of the public park in Hay El Jamea from a neglected area to a vibrant community hub stands as a testament to this.

The case of Hay El Jamea, therefore, enriches our understanding of placemaking in urban contexts. It advocates for a more inclusive, participatory approach that values the lived experiences and subjective perceptions of community members. This approach not only enhances the physical attributes of urban spaces but also imbues them with a deeper sense of authenticity and belonging. As such, the pursuit of meaningful urban environments necessitates a blend of bottom-up engagement and thoughtful, inclusive planning, paving the way for the creation of spaces that resonate with and fulfill the complex, varied needs of urban dwellers.

Funding: This work was supported by the Open Access Publishing Fund of the Free University of Bozen-Bolzano.

Informed Consent Statement: Informed consent was obtained from all subjects involved in the study.

Data Availability Statement: Data is available in a publicly accessible repository that does not issue DOIs. Publicly available datasets were analyzed in this study. This data can be found here: http://www.ghobeiry.gov.lb.

Acknowledgments: I would like to thank the Municipality of Ghobeiry for all their support during this research.

Conflicts of Interest: The author declares no conflicts of interest.

References

1. Totaforti, S. Emerging Biophilic Urbanism: The Value of the Human–Nature Relationship in the Urban Space. *Sustainability* **2020**, *12*, 5487. [CrossRef]
2. Lester, S. An Introduction to Phenomenological Research. 1999. Available online: https://devmts.org.uk/resmethy.pdf (accessed on 1 May 2023).
3. Tsai, B.-W.; Chung, M.-K.; Hsu, Y.-L. Exploration of the Meaning of a Place by Volunteered Geographic Information. *Asian J. Humanit. Soc. Stud.* **2018**, *6*, 4. [CrossRef]
4. Seamon, D.; Sowers, J. (Eds.) *Place and Placelessness, Edward Relph*; SAGE Publications Ltd.: London, UK, 2008. [CrossRef]
5. Norberg-Schulz, C. *Genius Loci: Towards a Phenomenology of Architecture*; Rizzoli: New York, NY, USA, 1980.
6. Jive´n, G.; Larkham, P.J. Sense of Place, Authenticity and Character: A Commentary. *J. Urban Des.* **2003**, *8*, 67–81. [CrossRef]
7. Musa, H.; Shok, M. The Role of Urban Memory in Reviving the Place (Al-Rasheed Street is a Case Study). *Iraqi J. Archit. Plan.* **2022**, *21*, 90–105. [CrossRef]
8. Sokolowski, R. *Introduction to Phenomenology*; Cambridge University Press: Cambridge, UK, 2000.
9. Merleau-Ponty, M. *Phenomenology of Perception*; Routledge & Kegan Paul: London, UK, 1962.

10. Casey, E.S. How to get from space to place in a fairly short stretch of time. In *Senses of Place*; School of American Research Press: Santa Fe, NM, USA, 1996; pp. 13–52.
11. Bachelard, G. *The Poetics of Space*; Presses Universitaires de France: Paris, France, 1958.
12. El Moussaoui, M. The Phenomenological Significance of Dwelling in Architecture. The Case of Eastern Beka'a Valley-Lebanon. Ph.D. Thesis, Universitat Politècnica de València, Valencia, Spain, 2020.
13. Heidegger, M. Building Dwelling Thinking. In *Poetry, Language, Thought*; Harper & Row: New York, NY, USA, 1971; pp. 145–161.
14. Wesener, A. 'This place feels authentic': Exploring experiences of authenticity of place in relation to the urban built environment in the Jewellery Quarter, Birmingham. *J. Urban Des.* **2015**, *21*, 67–83. [CrossRef]
15. Jackson, J.B. *A Sense of Place, a Sense of Time*; Yale University Press: New Haven, CT, USA, 1994.
16. Cilliers, E.J.; Timmermans, W.J. The Importance of Creative Participatory Planning in the Public Place-Making Process. *Environ. Plan. B Plan. Des.* **2014**, *41*, 413–429. [CrossRef]
17. Tuan, Y.-F. *Space and Place: The Perspective of Experience*; University of Minnesota Press: Minneapolis, MN, USA, 1977.
18. Huyssen, A. *Present Pasts: Urban Palimpsests and the Politics of Memory*; Stanford University Press: Stanford, CA, USA, 2003.
19. Nora, P. Between Memory and History: Les Lieux de Mémoire. *Representations* **1989**, *26*, 7–13. [CrossRef]
20. Benjamin, W. *Illuminations*; Harcourt, Brace & World: New York, NY, USA, 1968.
21. Healey, P. *Collaborative Planning: Shaping Places in Fragmented Societies*; UBC Press: Vancouver, BC, Canada, 1997; p. 288.
22. Lefebvre, H. *Writings on Cities*; Blackwell Publishers: Cambridge, MA, USA, 1996; 250p, ISBN 9780631191872.
23. Hayden, D. *The Power of Place: Urban Landscapes as Public History*; The MIT Press: Cambridge, MA, USA, 1995; pp. 292–295.
24. Traboulsi, F. *Loubnan al Hadith [Lebanon's Modern History]*; El-Rayyes Books: Beirut, Lebanon, 2008.
25. Tabet, J. *Beyrouth, Collection Portrait de Ville [Beirut: Portrait of a City]*; Institut Français d'Architecture: Paris, France, 2002.
26. Fawaz, M. Neoliberal urbanity and the right to the city: A view from Beirut's periphery. *Dev. Chang.* **2009**, *40*, 827–852. [CrossRef]

Disclaimer/Publisher's Note: The statements, opinions and data contained in all publications are solely those of the individual author(s) and contributor(s) and not of MDPI and/or the editor(s). MDPI and/or the editor(s) disclaim responsibility for any injury to people or property resulting from any ideas, methods, instructions or products referred to in the content.

 *architecture*

Article

A Controversial Make-Over of a 'Make-Believe' Heritage—The Transformation of Guangrenwang Temple

Lui Tam 

Welsh School of Architecture, Cardiff University, Cardiff CF10 3NB, UK; taml@cardiff.ac.uk

Abstract: This article discusses issues related to sustainable heritage management in China and problematises two dichotomies in heritage practices and research: the 'Eastern/Western' approaches and the tangible–intangible divide. It addresses these issues by examining the dramatic 'make-over' project of Guangrenwang Temple in Shanxi Province, China. The 'make-over' project transformed a small rural temple with a ninth-century timber structure into an architectural history museum, with a combination of private, public, and crowd-sourced funding. A real-estate corporation played a significant role in the project's initiative and organised a large-scale national and international publicity campaign around the project. Previously unknown to most laypeople in China, the temple attracted much debate since the project's completion, revolving around its 'cultural legitimacy', the design's appropriateness, the sustainability of the revitalisation, and the implications of the project to its 'heritage value' and authenticity. This article traces the opinions, actions, and effects of the temple's heritage assemblage and reveals the causal powers contributing to the emergence and transformation of associations within. It further questions the project team's claims regarding the project's effects on the historic setting's authenticity and its long-term social impact on the relationship between the temple and its community. It reveals five controversies regarding the choice of its curation theme, architectural language, decision-making, and management models. The complexities manifested in the actors' actions and effects demonstrate the ambiguous boundaries between the tangible and the intangible, and the perceived 'Western' and 'Eastern' approaches.

Keywords: sustainable heritage management; heritage museum; China; Eastern/Western approaches; tangible; intangible

Citation: Tam, L. A Controversial Make-Over of a 'Make-Believe' Heritage—The Transformation of Guangrenwang Temple. *Architecture* **2024**, *4*, 416–444. https://doi.org/10.3390/architecture4020023

Academic Editor: Johnathan Djabarouti

Received: 29 April 2024
Revised: 3 June 2024
Accepted: 4 June 2024
Published: 11 June 2024

Copyright: © 2024 by the author. Licensee MDPI, Basel, Switzerland. This article is an open access article distributed under the terms and conditions of the Creative Commons Attribution (CC BY) license (https://creativecommons.org/licenses/by/4.0/).

1. Introduction—Research Context

1.1. Two Intertwined Dichotomies

In the last four decades, heritage in China has experienced an unprecedented 'bloom', or even 'craze' and 'fever' (pp. 10–14, [1,2]). This bloom is demonstrated not only through its keen pursuit of international recognition through UNESCO but also the increasing attention given to cultural heritage domestically. In recent years, an expanding volume of academic literature on China's heritage phenomenon has brought critical reflections and multi-disciplinary perspectives to the subject area. This article continues these critical reflections and interdisciplinary enquiries into China's heritage practices, focusing on a controversial heritage transformation project in Southern Shanxi Province. The case study questions some assumptions in previous heritage research on China, including those from Critical Heritage Studies (CHS).

The 1994 Nara Document highlights the significance of contextualisation when assessing authenticity in heritage [3]. Since then, Western heritage practitioners and researchers have been keen to emphasise the Eastern/Western dichotomy in heritage approaches, partly based on a widespread misunderstanding and misconception of Japanese practices such as the Ise Shrine [4,5]. The critical turn of Heritage Studies in past decades has highlighted the conflicts between competing discourses in non-Western contexts, characterising them as a result of the distinctiveness between 'Western' and 'non-Western' traditions, cultures, religions, and ideologies [6–11]. However, the hasty acceptance of this dichotomy has led to overly simplified, essentialised, and even romanticised views of heritage activities in non-Western contexts [12–15]. Instead of the 'discourse of differences', termed by Winter [16] to describe and criticise the perceived distinctions between the 'materialistic Western approach' and the 'non-materialistic Eastern approach', the processes of negotiation between these various approaches can be better described as both differentiation and assimilation [12].

As conceptualised by Smith, the Western approach emphasises the monumentality and tangibility of heritage promoted by states and international organisations such as UNESCO [17]. However, the non-Western alternatives characterised in the literature have less universal definitions and are often less explored. Specific features of these alternatives are frequently used for over-generalisation rather than further discussing the complexities within (cf. [18]), a problem that scholars of Orientalism raised over two decades ago [19,20]. This simplified dichotomy also hinders a more nuanced understanding of the subject matter in Western contexts. The conflicts, negotiations, differences, and assimilation processes in each unique country or region deserve more nuanced and in-depth scrutiny [12].

Moreover, the 'East/West' dichotomy is closely intertwined with the conversation between heritage's tangible and intangible aspects, as they are often stereotypically assigned to one context or the other. However, the accuracy of claim that non-Western societies lack interest in tangibility and prefer intangible 'folkways' is very much debatable in the contemporary era (cf. [21]). Even though studies render the impracticality of separating the two [22,23], they are still widely used to categorise heritage entities [24,25].

1.2. Heritage in Post-Cultural Revolution China (Post-1978)

The development of China's cultural heritage industry is made possible not only through the resources brought by the economic growth since the 'Opening Up' in 1978 [26] but also partly due to the perceived 'threat' associated with the heritage sites lost, or potentially lost, to the process of the very same development. During the early stage of the economic reform in the 1980s, 'use first' rather than conservation was the heritage principle, which led to many heritage sites, even listed ones, being lost in the rapid economic development process [27]. The perceived 'threat' is one of the main characteristics of heritage policies and public discourse in post-Cultural Revolution (post-1978) China [1,28]. The urgency and the implied 'threat' that heritage faces are characterised as a notion tightly bound with modernity, which also gave rise to the heritage boom in the late modern period in the West [29]. This notion of 'threat' permeates China's heritage discourse, research, and practices, especially in the post-Cultural Revolution era. For Chinese intellectuals, such 'threats' are not only posed to Chinese heritage, but also, more broadly, to the 'Chineseness' contributing to the national identity [30].

Zhu and Maags ([2], p. 13]) contend that this 'heritage fever' is not only a state-led political move to strengthen the ruling party's power and reinforce a unified national identity narrative but also a trend experienced by the Chinese populace. An array of actors, including entrepreneurs, academics, and other individuals, participate in heritage activities enthusiastically and consume 'heritage products' with eagerness, often with active participation and collaboration from the local governments. The active engagement of academics in higher education institutions in heritage practices has blurred the division between heritage academics and professionals. The 'heritage bloom' also manifests as increased state funding for heritage projects and academic research (Figure 1). However, when the heritage 'fever' is coupled with the sense of urgency prompted by the perceived 'threats', it could lead to large-scale but short-sighted and compartmentalised 'rescuing missions' that do not consider their long-term impact and continuation.

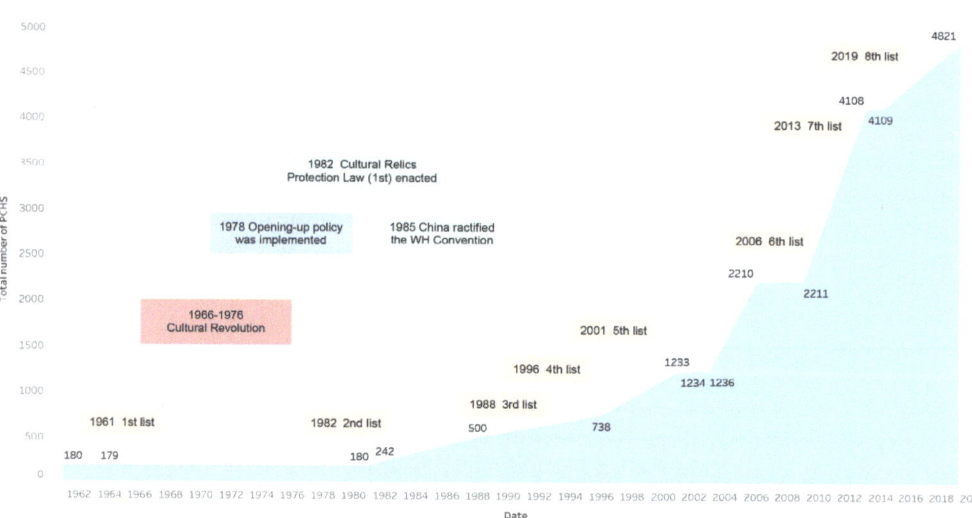

Figure 1. The number of national Protected Cultural Heritage Sites (PCHS) listed from 1960–present [31].

In recent Anglophone critical academic commentaries on China's heritage phenomena, particularly those from the CHS, two aspects of dissonance have been highlighted. One concerns the power relation between state and non-state actors [32–34]. There have been emerging yet insufficient discussions on how local actors can also play a part in negotiating with the dominant narrative and state-led Authorised Heritage Discourse (AHD) through individual agencies [35]. The other aspect concerns the negotiations between a supposedly imported 'Western' approach and the 'Chineseness' advocated by state actors and non-state actors [2,10,36–39].

The perceived distinctiveness of traditional Chinese philosophy from the Western AHD can be misleading. For example, the dynastic political interruptions in China's historical times are often cited as the reason for the lack of surviving historic buildings, demonstrating Chinese traditional society's lack of interest in preserving the historic built environment [2,40]. However, ample archaeological evidence shows that social organisations on a local level that maintained and restored historic buildings, including local communities and craftsmen, had shown keen consideration of preserving and reusing previous building materials and architectural forms [41,42]. A more fine-grained and nuanced characterisation of these aspects has long been needed.

Bridging and dissolving the two dichotomies present a research opportunity to fully capture the complexity and omnipresent interconnectedness of heritage's tangible and intangible aspects in a contemporary context. The case study in this article sets out to answer the questions: What does the case study demonstrate that can help overcome the Eastern/Western and tangible/intangible dichotomies? How can critical reflections on heritage approaches impact practices and decision-making? What fundamental mechanisms and conditions give rise to the (un)sustainable outcome in heritage management?

2. Theoretical Framework and Methodology

2.1. Theoretical Framework and Research Design

The case study reported here is part of broader research on sustainable heritage management in contemporary China. The research adopts a relational and dynamic framework inspired by Assemblage Theory (as developed in [43]) and Critical Realism (CR) [44–48], defining sustainable heritage as 'an assemblage sustained by human and non-human actors connected by enduring and dynamic associations' [49].[1] The framework examines how these associations and actors emerge and evolve through the assemblage's life cycles. It investigates what impact certain events or interventions might have had on the assemblage by scrutinising the tendency for change (or the lack thereof) in the associations and actors.

Case study research as a methodology guides the research design, allowing in-depth empirical investigation and analysis [50]. The broader research informing this article focuses on a case study region, within which three cases are selected for further scrutiny, one of which is reported in this article. This research design is particularly deployed to answer questions related to China's heritage management complexities and to identify the generative mechanisms for sustainable heritage management without any presumptions of which and how actors and mechanisms are at work. The incorporation of CR brings a focus on causality into the research and identifies the generative mechanisms through abstraction [51–53]. However, before this step, it is paramount to trace the causal chains of events to understand the causal powers at play.

Tracing the causal links within the broader research is achieved through several analytical steps: 'explanation building', 'time-series analysis', and 'cross-case synthesis' (pp. 212–255, [54]). This article will not elaborate on the third step, but the broader regional context will inform the arguments. The data analysis starts with a qualitative description for explanation building [55]. Controversy mapping [56] navigated among discourses, attitudes, and actions, elucidating the most contested issues within this complex case. Time-series analysis traces the life cycles of the actors and associations involved in the case study. It identifies the causal powers contributing to these associations' emergence and evolution.

2.2. Case Study Selection and Data Collection

The broader research informing this article focuses on the South and Southeast Shanxi region in China, with a high concentration of surviving early timber structures (pre-14th century). This research, conducted between 2017–2022, started upon the completion of a decade-long (2005–2015) national scheme, 'The Southern Project' (*Nanbu Gongcheng*), aimed at 'rescuing' and restoring 105 early timber structures in this region. The early timber buildings in this region are among the most typical heritage recognised by China's current administrative and legislative system and heritage professionals. The conservation and management approach to this heritage type shapes many of the approaches taken towards other types of heritage. Moreover, this research reveals many complexities and controversies, even in these 'typical' heritage sites.

The case studies were selected based on the presence of controversies and the accessibility of data. The Guangrenwang Temple case is exceptional in terms of these two criteria due to the extensive publication of actors' opinions, a record of public discourse and controversies documented on social media, the presence of a high-profile and heavily invested intervention with prominent external actors and demonstrated challenges in facilitating its sustainable future.

The data were obtained from documents, archival records, interviews, and direct observations. The documents and archival records include heritage records of the national PCHS published by the State Administration of Cultural Heritage (SACH)[2] and other statistical records from the census, national and provincial policies, and legislative and regulatory documents relevant to heritage conservation and management in China. The case-specific documents and records include the transcriptions of the historic stone steles and other inscriptions on site, historic and contemporary chronographies, project design drawings and documentation, administrative documents, news and magazine articles, and social media entries relevant to the case.

For the broader research, two fieldwork studies were conducted in 2018 in the case study region, the capital city of Shanxi Province, Taiyuan, and Beijing. Within the case study region, 53 national PCHS with early timber buildings across 16 counties were investigated. A total of 71 semi-structured focused interviews were conducted with six categories of actors, including (A) national officials in SACH and provincial-level officials in the Cultural Heritage Bureau of Shanxi Province; (B) local-level officials in heritage management and other relevant departments (municipal, district, and county levels); (C) on-site managers and caretakers of the PCHS; (D) local community members; (E) heritage professionals; and (F) local craftsmen and artisans. For the Guangrenwang Temple case, nine relevant participants from categories A, B, C, and E were interviewed directly. Due to this research's financial and temporal constraints, it was not possible to collect broader longitudinal data such as more community members' opinions over time, which is a limitation. To mitigate this limitation, the opinions of broader local community members, the architects and entrepreneurs involved, the general public, and other commentators were collected and recorded indirectly through publications, social media platforms, and other online records. Besides interviews, direct observations cover PCHS's physical conditions and settings and people's behaviours and interactions with the space, documented with field notes, photographs, videos, and mapping.

3. A Closer Look—The '*Long* Plan' at Guangrenwang Temple

3.1. Background—The Make-Believe Heritage

Guangrenwang Temple, also known as the Wulong Temple (Five-Dragon Temple), is in Zhonglongquan (lit. 'middle dragon spring') Village in Ruicheng County, Yuncheng Municipality (Figures 2 and 3). The temple is on an earth mound northeast of the village, about seven kilometres north of Ruicheng, in the northwest corner of the ancient Wei City ruins (ca. 403–225 BCE). About 800 metres southeast of Guangrenwang Temple is the well-known relocated Yuan Dynasty (1366–1468 CE) Yongle Taoist Temple [57].[3] Guangrenwang Temple was designated as a provincial PCHS in 1965 and a national PCHS in 2001. Only two historic buildings are left on the temple ground [58]. The north-facing stage is located on the south end of the mound, opposite the south-facing main hall. The surviving stage was constructed in the Qing Dynasty (1644–1911 CE) while the main hall is believed to have retained a Tang Dynasty (618–907 CE) timber structure. Two side halls were described to have once been on the temple ground [59].

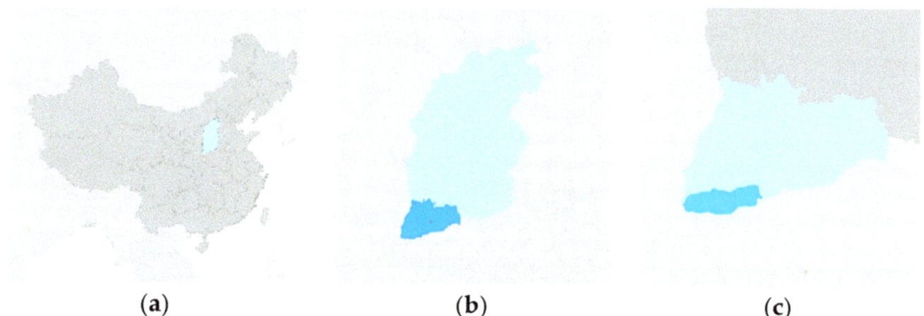

(a)   (b)   (c)

Figure 2. (a) Shanxi Province in China; (b) Yuncheng Municipality in Shanxi Province; (c) Ruicheng County in Yuncheng Municipality, base map GIS data: [60].

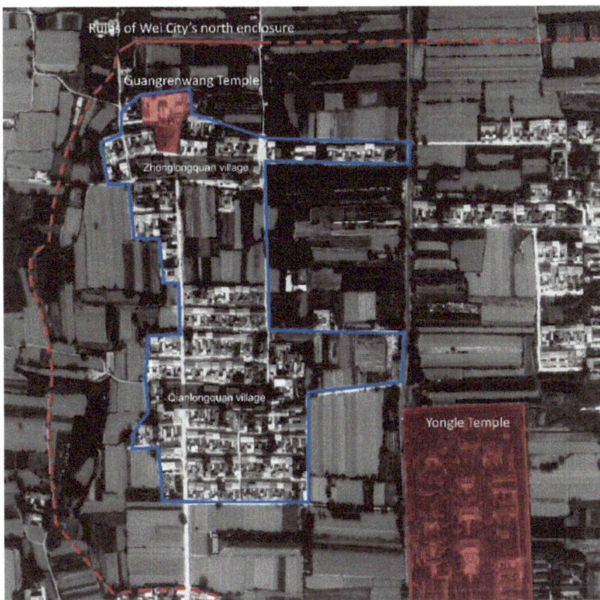

Figure 3. Location of Guangrenwang Temple in relation to Zhonglongquan village and Qianlongquan village, Yongle Temple and the ancient Wei City, base map: [61], annotated by the author.

On the lower ground, in front of the temple, was a pond, referenced in historical records as the Dragon Spring, closely related to the temple. The pond has been dried out for decades due to the decreasing underground water level, but its shape is still visible. Across from the fields north of the Wei City site, Zhongtiao Mountain forms the backdrop of the temple, another crucial element of its natural setting [62,63]. The earliest known academic literature hypothesising that the main hall was a possible Tang structure was published in 1959 [58]. Jiu asserted that despite having experienced several alterations after its initial construction, the architectural structure of the main hall still retained the 'Tang Dynasty style' (ibid. p. 43). Several other architectural historians subsequently endorsed this view, which became a consensus [64–66]. The verdict is essential to reading the 'heritagisation' of the temple as one of the country's four surviving Tang Dynasty timber structures.[4]

After it was 're-discovered' in 1958, a restoration was carried out in the same year by the Shanxi Cultural Relics Management Committee and Ruicheng County People's Committee.[5] Subsequently, SACH issued criticism regarding the restoration's intrusive interventions [67], but what percentage of the structure has been 'altered' has never been assessed. Chai, a renowned architectural historian, also criticised the fact that many of the components have been replaced without a basis of sound historical evidence [57]. Chai considered that it would be advisable to 'restore the structure to follow the Tang style' in the future [57], which suggests that his criticism towards the 1958 restoration focused more on the incorrect form of the components used for replacement rather than a concern that the historic fabric was substantially replaced.

Despite contested opinions over whether the main hall can still be qualified as a 'Tang Dynasty structure' due to the 'heavy-handed' restoration in 1959, the 'belief' that this temple is of outstanding significance for containing one of the few surviving Tang structures in the country remains persistent [67]. It should be emphasised here that the Tang structure's materiality *is* essential to the representational 'heritagisation' process on a national level. However, the 'material authenticity' seems to have become less vital after it became PCHS. This 'make-believe' mentality was instrumental and strengthened during the 2016 'make-over' project that followed the 2015 restoration.

The 1959 article states that the dragon king statue was missing then. No information has been found regarding when the temple lost its historic statues and other temple buildings. Jiu also recorded that after the 1958 restoration, the temple was a 'recreational venue for the people' and a scenic attraction [58]. It suggests that the temple's religious status was somehow considered as being in the past, and the desire to use it for tourism was already present in the 1950s. Guangrenwang Temple was used as a primary school during the 'Socialist Transformation Movement' in the 1960s–1970s. The primary school moved away in 1981, and the temple was never registered as an official religious venue under the new administrative system for religious affairs [67,68]. Nevertheless, statues were re-installed. Even though the temple might not have the same popularity as a space of worship as it used to, there are still religious activities in the temple since its heritagisation.

Despite its proximity to the famous Yongle Temple, Guangrenwang Temple has rarely received visitors since the 1950s. The temple, especially its surroundings, slowly became dilapidated after the primary school left [67]. Before the 2013–15 restoration, the temple was only enclosed by a short brick wall and a small picket fence that could be easily breached. There have been caretakers at the temple since 1993. However, their presence was far from enough to safeguard the temple site. In December 2012, one of the two Tang steles was stolen, which, fortunately, was soon recovered [67].

3.2. The 'Long (Dragon) Plan'—The Post-Restoration Make-Over

Such was the temple's situation when Ding, the senior vice president of Vanke, one of the country's largest residential real estate developers, allegedly went on a historical architectural tour in Shanxi and encountered Guangrenwang Temple in 2012. Subsequently, Vanke contacted the local authority of Ruicheng County, intending to get involved in the temple's restoration that was due to start in 2013. However, the local authorities in Shanxi and SACH were cautious and sceptical when a private corporation, especially a real estate developer, often perceived to be the 'enemy' of historic buildings, attempted to get involved in a heritage project. The authorities refused Vanke's request, citing that a national PCHS's restoration must be carried out by organisations with first-class qualifications in heritage conservation [69,70].

The restoration of Guangrenwang Temple under the Southern Project took place from 2013 to 2015 (Figure 4). An 'environment improvement' project was expected after the restoration. Such improvement projects are usually small and less significant than restorations [71]. Therefore, it was unusual when Vanke, a real estate company, announced to the public in June 2015 that a crowd-sourcing heritage project called the '*Long* (Dragon) Plan' had been set up to facilitate the environmental improvement of the temple to create a museum [72].

Figure 4. The main hall of Guangrenwang Temple after restoration, March 2018 (source: author).

Vanke associated the 'Long (Dragon) Plan' ('the Project' from hereafter) with Vanke's pavilion at the Milan Expo 2015 to boost international publicity. The pavilion, designed by Daniel Libeskind, was shaped like a dragon, covered with 4000 red tiles symbolising dragon scales.[6] To fulfil the Project's crowdfunding claim, Vanke auctioned the 4000 tiles to raise some of the funds from the public. The Project was eventually carried out with crowdfunding, private funding (from Vanke), and state funding [69].

URBANUS, an architectural studio based in Beijing, led by a star architect Wang Hui, was responsible for the design. According to Wang, there were two main objectives when they were conceiving the Project: to create an open-air exhibition space on the temple ground for Chinese architectural history and the temple's interpretation, and to create a public space for the villagers [73]. The Project was also supported by a team of heritage professionals from Tsinghua University, led by Lv Zhou. As a professor from Tsinghua University, the director of Tsinghua National Heritage Centre, and the vice president of ICOMOS-China, Lv's voice carried much weight and radiated influence among the administrative branches, academics, and the public. According to the Southern Project's leading engineer, who works extensively with SACH, the involvement and endorsement of Lv's team provided the provocative Project proposal with credit during the administrative approval process [74]. (See Figures 5–8 for comparisons of before and after the Project, and Appendix A for a detailed description of the Project's interventions on site, annotated in Figure 9).

Figure 5. Entrance to the temple via a slope before the Project, March 2011 (source: author).

Figure 6. New entrance with flights of steps towards the ticket office at the southeast corner of the museum, March 2018 (source: author).

Figure 7. Limited view of the main hall's roof from the top of the entrance staircase, March 2018 (source: author).

Figure 8. View of the main hall at the top of the entrance slope, March 2011 (source: author).

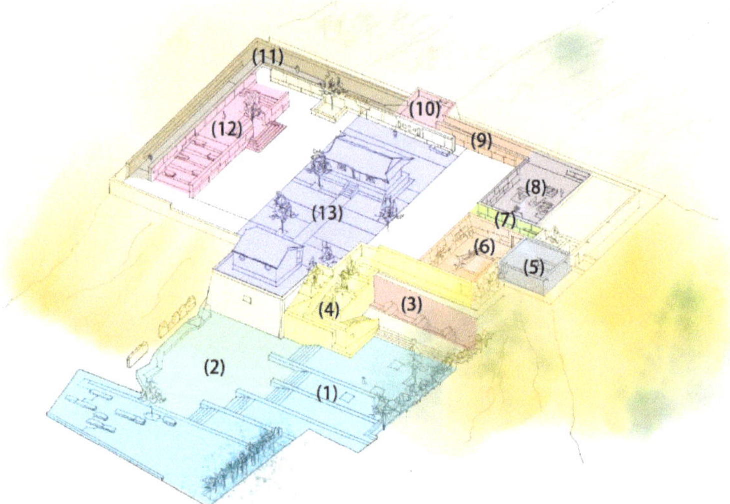

Figure 9. Axonometric view of Guangrenwang Temple after the Project [73], redrawn and annotated by the author (see Appendix A for a detailed description of each numbered area).

3.3. The Voices and the Silence—Five Controversies

The action of Vanke is characterised in most media reports as an act of philanthropy [69,75–77]. They tend to depict the initiative as stemming from individuals' spontaneous enthusiasm instead of Vanke's corporate action, and praise the fact that the chief architect took on the design pro bono [69,72,78]. However, it is equally reasonable to believe that Vanke chose this specific site because of its title as a Tang Dynasty structure. It was important enough to create good publicity but also sufficiently 'unknown' to the public that a case of 'rescuing' and 'revitalising' could be made [79]. A closer look at the debates and reality of the case suggests that behind the positive media coverage, there are many controversies worth unfolding in order to appraise the Project critically.

The heated nation-wide debate started when the outcome of the transformation was revealed in May 2016. Many were surprised to see the images of the museum publicised online, which received various comments from the public. Furthermore, heritage professionals, architects, academics in relevant disciplines, and the media started publishing more detailed responses to the Project (Figure 10). Subsequently, the chief architect and heritage professionals involved in the Project responded to some of the most contentious issues.

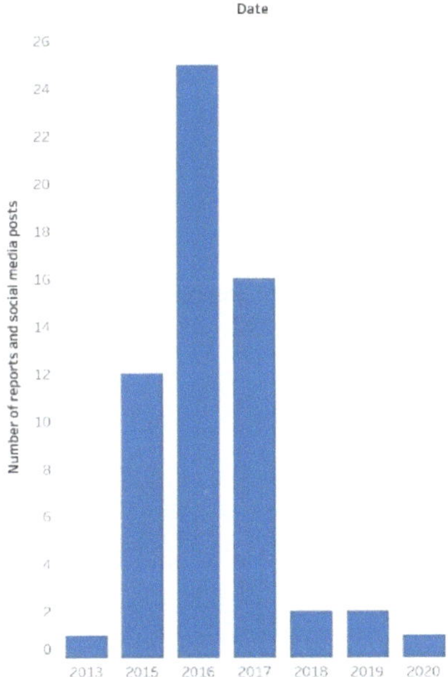

Figure 10. The number of academic articles, op-eds, media reports, and social media posts about the Project before and after its completion; data collected by March 2020.

Data collected during this research show that the actors' opinions had changed over time. For example, according to Ruicheng's local officials and a media report, when the provincial officials first came to see the site immediately after its completion, they were hesitant to voice their opinions [80]. Such hesitation was no longer evident during the interviews in 2018. Changes were also present in the caretakers' opinions from different times. According to media reports, the caretakers were optimistic about the change upon the Project's completion [67,78]. However, in the 2018 interview, one of the caretakers voiced detailed complaints about the Project's construction quality and design choices. They were also dissatisfied that the caretakers and community members were not consulted until the construction stage [81]. This discovery demonstrates that changing opinions should be considered when assessing the Project's long-term impact.

3.3.1. Controversy I—The Authenticity of the Historic Setting

Commentators hold contrasting opinions regarding the Project's impact on the authenticity of the historic setting (*lishi huanjing*). These debates not only demonstrate the actors' opinions but also their understanding of authenticity as a concept. Architects, heritage professionals, and architecture magazine commentators tend to refer to the term explicitly. Within these debates, the authenticity of the site's historic setting involves both the physical setting and the anthropological and socio-cultural contexts.

Wang, the lead architect, claims that the Project did not intervene with the temple's historic setting because there was no material evidence remaining, citing that the site's physical environment has been altered several times over the centuries [73]. Following the same reasoning, Huang suggests that the heritage museum is, therefore, only yet another layer of this ever-changing setting [82]. Lv (also appearing as Lyu) and Guo comment from the perspective of heritage professionals. They argue that it is reasonable and 'respectful to history' to create something modern instead of a 'reconstruction out of imagination' due to a lack of tangible trace left of the 'historic layout' [83,84]. However, citing the Venice Charter and the Nairobi Recommendation, Guo admits that the new environment does not support the role of the historic temple as a 'testimony of history' [84].

Conversely, several commentators suggest that the Project is not sensitive to the broader physical setting—the rural village and surroundings [74,85]. Qi and Li note that the Dragon Spring Pond, the surrounding fields, and mountains are essential elements of the temple's historic setting, evidenced by the historical records in local chronographies and historic steles on-site [62,63]. Qi and other media outlets report that the caretakers regret the design team's decision not to recover the Dragon Spring fully, aligning with the interview with the caretaker during this research [67,78,81].

The lead engineer of the temple's latest restoration interprets its authenticity based on the philosophical position of the ICOMOS Principles for the Conservation of Heritage Sites in China (the 'China Principles' hereafter), where minimal intervention is the overarching principle [86]. They consider that heritage professionals are responsible for interpreting and promoting the China Principles. After trying to minimise the intervention when restoring the main hall, they are especially disappointed to see the drastic transformation of its environment [74].

In response to such criticism, Wang references the minimum intervention principle from the Venice Charter and claims that the design has been revised multiple times to avoid unnecessary interventions with the purpose of 'keeping the heritage in its setting with authenticity and integrity' (p. 113, [73]). However, Wang goes on to argue that keeping the authenticity of the setting is paradoxical in practice because the archaeological survey did not find any physical evidence of the historic setting. Therefore, there is no reliable source from which to draw a hypothesis. This argument shows that Wang's definition of authenticity has changed slightly from when he references the Venice Charter. In the latter instance, he understands 'retaining authenticity' as a requirement to reconstruct the temple's historic layout.

The way visitors approach the main hall is one of the most debated design decisions. Chinese historic buildings are mainly approached from their front façades. Such an approach accentuates the significance of the temple's main hall. The spatial relationship between the main hall and the north-facing stage is the most essential element of the layout. It is the connection between the deity worshipped in the main hall and the people who perform for the spirits on the stage. This association, symbolising the relationship between humans and the place, has existed since the creation of this temple.

The spatial organisation of the museum now takes visitors from the entrance, where the main hall is blocked from view, to a narrow opening towards the side façade of the main hall flanked by two walls. The architect is indeed very proud of this framed image as it appeared in his original design sketches, and it is most often photographed after the Project (Figures 9 and 11) [87]. By blocking the visitors' peripheral vision to achieve a framed side-elevation of the main hall, the design has distracted the visitors from discovering the connection between the stage and the main hall. As revealed from the architect's argument, the decision-making was primarily based on one attribute of the site's heritage value—its status as a Tang structure [73]. The choice to highlight the single authorised discourse and view the main hall as a stand-alone 'large museum object' appears deliberate [87]. The de-contextualisation of the main hall is praised by some architects such as Zhou and Lu, who consider its museumification as a post-modern way to present the building [87,88].

Figure 11. Elevation of the main hall from the corridor towards the temple ground (source: author).

Chen, however, is critical towards the neglect of the association between the main hall and the stage. As a heritage professional, she perceives this association as the most significant testimony of its heritage value (being a local temple) and should be respected by any project [85]. Furthermore, Liu notes that the relationships among the village, the pond, the temple, and the field are the most significant associations relevant to the Project, as they symbolise the associations among heaven, earth, and people. He suggests that the temple's creation and the century-old worshipping activities have fostered a shared meaning among the local population. He argues that such a shared meaning should be preserved and enhanced by the Project, which, in his opinion, has not been achieved [85].The comments on authenticity also engage with the intangible setting—the socio-cultural 'landscape' of the community. Guo considers that the new heritage museum diversifies the connections between the historic temple and its audience, as a museum of ancient architecture, a heritage site, and a temple. He concludes that the Project positively impacts the 'reconstruction' of the intangible setting [84]. Conversely, the Southern Project's leading engineer comments that the heritage site now no longer looks like a temple created to serve the village and the local region. They argue that the Project has an adverse impact on the temple's historic function, which, according to them, constitutes part of its authentic historic setting [74].

In conclusion, by tracing the opinions and their underpinning philosophical positions around the concept of 'authenticity', this controversy demonstrates the relativity and contradictions in interpreting this concept often used by heritage commentators and professionals. Wang claims that the design's consideration regarding authenticity is demonstrated by transforming the site in a 'subtle' way, which has a positive impact on its authenticity through reinstating the connection between the site and the community and 'return[ing] kinship and folklore culture to the everyday life of the villagers' (p. 113, [73]). However, as discussed below, the choice of the exhibition theme and the decision to ignore the relatable association between the main hall and the stage brings this claim into question.

3.3.2. Controversy II—Space of Worship vs. Place of Knowledge

Another central issue refers to the dual associations of the site being a space of worship, built for a local population to worship a local deity, and a place of knowledge that came with its heritagisation, recognising its value as a testimony to Chinese architectural history. Wang and other commentators, including architectural and heritage professionals and journalists, base their arguments on the premise of the 'former' temple's spiritual obsolescence [67,73,83,87–89].[7] These commentators define the site's architectural significance as the embodiment of knowledge and consider folk culture and religions as the past, implying that the religious and folk connotations are not to be identified as 'knowledge'.

Such a judgement is biased against the grassroots meaning-making process around the temple that still exists today.[8]

Chen considers the site's most significant identity to be a temple created for worshipping a local deity. They suggest that this heritage value has not been sufficiently explored by the Project team [85]. Qi considers it false to presume that the historic temple is no longer a sacred space. Before the Project, the temple, although without an official religious venue registration, was still a space of worship on at least two occasions each month and during the temple fair related to Guanyu, a deity popular among the local population. Qi emphasises that this pattern of worship fits into the rhythm of everyday life in rural villages. She criticises the fact that the architect assumed responsibility for reinventing a connection between the site and the community out of imagination instead of respecting and enhancing what has existed and is still existing on-site [63].

Wang recalls that their team had considered two exhibition themes based on the value assessment published by SACH[9], which, he relates, were its architectural significance and the folk custom of rain-praying to the Dragon King. He admits that the final decision to exclude the latter was mainly based on the team's limited capacity, consisting primarily of architects who did not possess sufficient knowledge of folk cultures and religions [73]. This admission is telling. It reveals that the determining factor for this significant decision was the design team's limited specialism. It is then unsurprising that not only did the design team exclude the rain-praying custom, but they also overlooked the contemporary meaning-making process in the temple demonstrated by the local community's faith and worshipping activities.[10]

It should be noted that when all the interventions are considered, the spiritual connection has not been entirely ignored. New statues of the deities were installed in the main hall by Vanke. The local population has not stopped considering the site as a space of worship either [63,67,78,81]. It is, however, peculiar that this aspect is rarely mentioned in the initiators' responses to criticism of the Project. Similar silence can be found among the local authorities. Religion is still a politically sensitive topic for local officials, even though they all acknowledge the existence and significance of faith among the local population.[11]

3.3.3. Controversy III—The Design Language and Exhibition Curation

The Project's improvement of Guangrenwang Temple's physical environment is undeniable, considering that an informal landfill previously surrounded the historic temple. The controversy presented here focuses more on the effects of the design and museum curation on the site as a heritage space and a historic temple. According to media coverage, the public sector's opinions about the Project are mostly positive [67,69,72,75]. The administrations' reservations regarding the appropriateness of the design in written form can only be seen in the advisory and approval documents published by SACH [90,91]. However, interviews with officials from various administrative levels during this research reveal a less homogeneous set of opinions. Several local-level officials, including those not from Ruicheng County, express that they consider the design of the open-air museum to be 'too much'—too provocative and incompatible with the rural village environment [92,93].[12]

Li and Liu comment that the spatial organisation and volume of the added construction are disrespectful to the historic buildings as they made the main hall, which is not a large building, seem even smaller in the complex [85]. Additionally, Liu remarks that the circulation appears to guide him away from the main hall during his visit. This remark demonstrates the design's unintended impact of taking away the visitors' attention to the temple ground [85]. Relevant to this aspect, Guo and Zhou criticise the fact that the exhibition curation has abstracted the main hall into large-scale architectural drawings, purposefully framing the historic building into static elevations and creating a strict circulation route to access the main hall (Figures 11 and 12). In doing so, the design has taken away visitors' freedom to intuitively explore and experience the historic temple [84,87].

Figure 12. Introduction space with a section of the main hall and timeline display, March 2018 (source: author).

Nevertheless, Lv comments that using modern architectural language effectively distinguishes the new design from the historic remains of the temple (p. 2, [85]). Liu praises the fact that the unconventional entrance and the maze-like circulation are modern architectural languages but resemble classical Chinese gardens (pp. 6–7, [85]).

In his criticism of Liebeskind's Jewish Museum's cultivation of 'reductive approaches to built space', Koepnik considers places as articulations whose 'identities exceed the work of abstract and unified interpretations'. He emphasises the diversity of space and its association with history, memory, narratives, and uses (p. 346, [94]). The Project's interventions appear to have introduced a similar reductive approach. Behind the orchestrated architectural language and the curation theme is the explicit intention to create a museum as a top-down educational space. It is then unsurprising that the voices of villagers, whom the architect considers as needing to be 'educated', were not heard during the decision-making process. This intention is also related to one of the claimed achievements of the Project—the social impact, which will be discussed next.

3.3.4. Controversy IV—Social Impact of Public Engagement and 'Giving It Back to the Community'

According to Vanke, the Project was set out with three main objectives, the first of which was 'returning the temple to the village'. He explains that this objective derives from the hope to reinstate the temple's status as a public space in the village, to introduce the temple back into the everyday life of the villagers, and to encourage the villagers to care for the temple (p. 115, [95]). Wang, on the other hand, considers that in today's villages, virtues and faith are at a loss, and Longquan village is 'fortunate' to have this ancient temple. By 'returning the temple to the everyday life of the village', Wang appraises the fact that the Project is a 'redemption to the problematic village life' (p. 113, [73]).

Many commentators address the issue of public engagement and the Project's benefit to the village. According to Lv, one of the Project's most crucial contributions is the participation of various sectors of society [83]. This argument addresses a long-lasting condition of heritage practice in China, which had been limited to the public sector and government-appointed professional institutions. It was also why the administration rejected Vanke's first initiative to participate in the temple's restoration [69]. However, heritage professionals and some of the high-ranking government officials in the cultural heritage departments have been advocating for the participation of a broader range of actors from society in heritage projects, especially regarding adaptive reuse [96–108]. The eventual acceptance of Vanke's second initiative demonstrates that the public sector has become more open to private sector involvement.[13]

Lv praises the touching fact that after the completion of the Project, villagers, elderly people, and children were seen enjoying the site [83] (Figure 13). However, based on the observation in 2018 and interviews with Ruicheng's local officials and the temple's caretaker, the frequency of the villagers' visits to the site has only moderately increased, even though they do use the public square.[14] The increase of non-local visitors is also insignificant after the initial excitement. According to the local official, the annual income from the ticket fees was about 20,000 CNY (approx. 2200 GBP) in 2017. It was better than before the Project when there was no income at all, yet it was hardly enough for the maintenance of the site, let alone to bring any extra benefit to the village [80]. The impact (or the lack of it) of the Project on the local development opportunities, which is one of the claims of the Project's initiators, will be further discussed in Section 3.3.5.

Figure 13. The public square created by the Project in front of Guangrenwang Temple, March 2018 (source: author).

As pointed out by Zhuang, the Project's decision-making, either regarding the spatial organisation or the exhibition theme, received no input from the community [85].[15] According to the temple's caretaker, he could only voice his opinions when the construction started, and the architects altered part of the plan accordingly [81]. The lack of community participation reflects not only the project team's lack of awareness but also an institutional flaw embedded in the current administrative procedures of heritage projects in China. As a project of a national PCHS, the proposal only needs to be appraised by heritage professionals and approved by the various levels of administration, who are more concerned about abiding by the heritage regulations and legislation than whether the project meets the local population's needs [90,91].

In conclusion, although the initiators and the architect claim that the Project has harnessed significant benefits for the local community by inserting a museum in a rural setting, evidence indicates a lack of impact on the villagers' willingness to visit the museum and the local economy. While the museum management has returned to the local administrative department, the villagers' participation in the Project or the museum's management is minimum. While it is undeniable that the Project has enhanced the 'social values' of the site by allowing private sector participation and attracting public attention, the short-term nature of the Project also means that the private sector investment is probably a one-off incident, and public attention quickly died out after 2016.

3.3.5. Controversy V—Revitalisation and Sustainable Management

All the above controversies point to an overarching issue regarding the sustainability of the Project and the heritage assemblage of Guangrenwang Temple. Ding, the initiator from Vanke, praises the project's objective to transform 'dead heritage' into 'living heritage', echoed by Zhou (p. 1, [85,87]).[16] Ding believes the Project can potentially provide a model for revitalising other local heritage sites [85]. Such a belief is shared by local and provincial officials, who are hoping to promote this model in Shanxi Province [80,109].

Compared to restoration projects, cultural heritage administrations are much more relaxed regarding environmental improvement projects because they do not directly involve the PCHS's historic structures. It is also more desirable for the provincial and local authorities for them to be privately funded due to a decrease in state funding for such projects since 2015 [69].[17] The Project's eventual approval by SACH and the praise from the public sector show that there has been a desire to 'think outside of the box' within the administration. Such a project would have been almost unthinkable just a few years before. Nine interviewees, including heritage professionals and provincial and local officials, who have commented on the Project, all agree that it was positive in attracting attention from the broader public and getting more actors in society involved in heritage management [74,79,80,92,93,109–113].[18] However, while finding a suitable new function for a heritage site might be a good start, sustainable management afterwards is often very challenging.[19] This concern is aligned with the current situation of the museum at Guangrenwang Temple. Two years after the project's completion, the revitalisation's impact had already become questionable. The lack of planning for the museum's sustainable management might eventually make the effort futile.[20]

According to the architect Wang, maintaining a national PCHS like this cannot be supported only by a remote village but needs to attract tourism. It indicates that one of the design's objectives is to attract an external audience. Hou from Vanke also states that the Project has attracted more visitors and increased the village's income. He considers that such a change will create new opportunities for a traditional village like Longquan [95]. However, based on the information gathered during the fieldwork of this research, such a claim appears questionable.

According to the village chief of Longquan village, the entrance fee income from the museum is first submitted to the centralised county financial system, and the same amount is returned to the heritage site for its maintenance and management [67]. Therefore, the income of the heritage museum does not directly benefit the villagers. Furthermore, according to the village chief and a local community member who took part in the Project as a project supervisor, since there is no other supporting infrastructure or other attractions in the area, the mode of visiting is unlikely to change despite the potential increase in visitor numbers. Currently, tourist visits usually involve a two-hour to half-day tour of Yongle Temple and Guangrenwang Temple in Ruicheng [67]. By 2018, according to local officials in Ruicheng County, there was a plan to sign off a contract for a private company to develop rural tourism in the village. The company will not be allowed to do anything more with the temple, and there is no further plan regarding how the open-air museum will be managed differently [80]. However provocative, the Project did not fundamentally transform the nature of the site's management. It is then predictable that enthusiasm and motivation to keep the site alive may not last long.[21]

Moreover, whether developing tourism with a heritage museum is suitable for a village like this is debatable. Literature in tourism studies has addressed contested issues regarding the role of community members in the decision-making processes, their share of benefits from tourism development, gentrification, and displacement of local settlements [114,115]. The 'gaze' of affluent urban tourists on the disadvantaged and low-income communities, exacerbating the inequalities between the consumers and producers in tourism development, has also been fiercely problematised [116–118].

The Project has created a museum space that requires more than grassroots efforts to maintain, let alone to update and renew the exhibition. The permanent exhibition facilities—the large bracket-sets models, the full-scale engraved architectural drawings, and information panels—make it difficult to add different exhibition themes to the museum to become more inclusive of local folk culture (Figures 14 and 15). This sub-section demonstrates that the Project's decision-making process predetermined its short-term effect. Indeed, Lv admits that despite the project's great potential for fostering opportunities locally, continuous observation is needed to determine how sustainable its management will be [110]. As with most heritage projects in China, the Long Plan was set out to be a one-off construction project. There was never a plan for grounded and long-term research to support the design or implement a sustainable management mechanism that returns a sense of ownership to the community. The decision of the exhibition theme strengthens the authorised discourse of the temple as a national PCHS. However, it ignores the other associations that are also present in the temple's history and more relatable to the local community, such as the religious connotation and folklore culture. Even if the community members are proud of the temple holding one of the few surviving Tang structures, a one-sided story about its architectural history does not excite long-lasting interest in their everyday lives. The lack of long-term engagement fails to sustain the associations between the temple and its local community.

Figure 14. Exhibition space with *dougong* models and information panels (source: author).

Figure 15. Corridor with information panels about early timber structures in Shanxi Province (source: author).

4. Discussion—Analysis of the Heritage Assemblage and Causal Powers

Like many other temples in the region, Guangrenwang Temple's assemblage started with the emergence of the religious association between the temple and its local community. The surviving historic documentation reveals that the temple's religious connotation is also related to specific associations between other actors in the local area, such as those between the temple and the Dragon Spring, between the Dragon Spring and the villages, and between the temple and the ancient Wei city. The drought seasons were explicitly recorded as causes for the construction and reconstruction of the temple. These associations, actors, the initiative of the local officials, and the specific climate conditions formed part of the generative mechanisms for Guangrenwang Temple's (re)constructions in the ninth century. Also present, albeit less explicitly expressed, was the local community's economic capacity to obtain the necessary resources to maintain the religious associations. The records on the historic steles also suggest a human–nature relationship between the local commune and the trees planted on site, which were both the consequence of and the resources for maintaining the religious association between the temple and the community.

The state-wide ideological movement in the mid-20th century significantly impacted the temple's religious association. The 'Socialist Transformation Movement' in the 1950s–1960s was the most apparent cause for such changes. However, the fact that Guangrenwang Temple no longer had worshippers by the 1950s suggests that broader social changes and the communities' dwindling religious faith at the time also played a role. The composition of actors who facilitated and participated in the 'Socialist Transformation Movement' is complex, including people and organisations from the state to the local levels, driven by state policies and ideologies. As in many other cases in the region, these mechanisms disrupted the historic religious associations but also created new associations between the temples' physical space and the community members who studied or worked there when the temple became a local school. The causal powers that started and sustained the 'Socialist' association were impactful and dictated by external actors. During this period, this association was the only prevailing one, as all other associations related to its religious function were subdued. However, in Guangrenwang Temple, this 'Socialist' association was short-lived and became obsolete as soon as the primary causal power coming from the state's ideological movement ceased to exist.

Another prominent association came into play as the heritagisation process of these sites started. For the sites studied in this research, the causal powers for their heritagisation almost exclusively came from recognising their architectural significance and the established administrative and legal system for heritage conservation in the country. The crucial role of this recognition in their heritagisation is particularly apparent in Guangrenwang Temple's case, as it was considered 'heritage' by architectural historians even before its early designation in 1965. Like many other sites in the region, Guangrenwang Temple was in a severely dilapidated state when it was rediscovered. The rediscoveries of these sites and their heritagisation, especially their designation as national PCHS, prevented them from complete physical obsolescence. In this sense, the site itself was a 'marginalised' actor in society, with its historic religious association with the local community dwindling or already obsolete. Heritagisation created a new association between the historic temples and broader society.

This association was strengthened further by the restoration under the Southern Project scheme from 2005–2015. The 'Long Plan' Project, a unique case in the region, triggered changes in multiple associations. Most notably, it strengthened its heritage association based upon its architectural significance even further after the latest restoration, bringing more of the general public to forge a new association with the temple as a heritage museum through public debates and publicity campaigns from Vanke. However, these connections, emerging through the heat of the topic, are weakened quickly as the public attention dies down. The limited increase in visitor numbers, the unchanged local management model, and sporadic activities organised by Vanke indicate that the association derived from its museum function will be challenging to sustain.

Despite the significant improvement in the temple's environment, the single focus on its architectural significance in the museum and the lack of input from the local community have significantly reduced the community's chance to forge any new association. The Project team's misconception of the temple's religious obsolescence further weakened the existing association with the local community derived from its religious connotation. Despite the good intention to create a public square for the villagers to socialise and 'return the temple to the community', the rest of the design has missed the opportunity to create an inviting environment for the villagers. At present, the heritage and museum associations are dominant among all other associations. While the heritage association was strengthened through the museum project and will likely be sustained, the museum association faces severe challenges in gathering enough causal powers for it to be sustained.

Comparing Guangrenwang Temple to another case, the Dongyi Longwang Temple in this region, where the community initiative and engagement are front and centre in its community museum's emergence (see [119]), reveals two opposite approaches. The themes of the two museums speak truth to this contrast, with the one in Dongyi Village proactively engaging the local culture and community and the one in Guangrenwang Temple deliberately excluding them. Despite the much smaller scale of investment and public attention, the Dongyi Village community is free and willing to pace the process according to their capacity and resources. Without the high-profile professionals and academics involved in the Guangrenwang Temple project, the community-based exhibition in Longwang Temple features more multidisciplinary involvement. On the other hand, it should be acknowledged that if community initiatives in the region can receive such generous financial support as the Long Plan project, it would provide more flexibility and opportunities for the grassroots activities. The crucial point here is not that powerful external actors like Vanke cannot be involved, but that such projects need to be more inclusive of the heritage assemblages' tangible and intangible elements of the heritage assemblage with a holistic approach.

For the cases investigated in the case study region, the state or provincial level policies, the religious faith of local communities, and the local initiative and capacity exercise crucial causal powers for sustaining the religious and heritage associations. A provincial official mentioned that Guangrenwang Temple's project might well provide a precedent for implementing the provincial scheme, the 'Safeguarding the Civilisation' Scheme [109]. This scheme would influence how similar heritage sites might be used or adaptively reused in the case study region. However, the Guangrenwang Temple case needs to be considered as a lesson learnt rather than a desirable outcome where the sustainable future of these sites is concerned.

The missed opportunities in Guangrenwang Temple's case to create or strengthen the associations between the local community, the site, and those among community members demonstrate that the connection between the heritage assemblage and actors in broader society is not guaranteed. The weakening museum association will afford fewer and fewer causal powers to initiate change in the local area, such as tourism development and other economic opportunities. Without sustained connections that bond the local community together, there is less opportunity for community-level initiatives, crowdsourcing, and collective actions to initiate change or sustain the museum in the future. The sustainable management of the temple as a religious venue, heritage site, or museum is relevant to the local community's potential to facilitate sustainable development. However, continued interactions between the sites and the local communities, either through religious worshipping or community engagement in the museum or heritage site, are crucial for this relevance to be translated into change.

Finally, the Guangrenwang Temple case suggests that new sustained associations would not automatically emerge from new adaptive reuse without the sustained and continuous involvement of actors. It also reveals that when an assemblage is dominated by a specific association while excluding others, it affects the entire assemblage's resilience and leaves little room for new associations to emerge. Conversely, a heritage assemblage with more sustained and coexisting associations reinforcing each other is more likely to be sustainable.

5. Conclusions

Through this case study, I have questioned the project team's claims regarding the Project's effects on the historic setting's authenticity and its social impact on the relationship between the temple and its community. The article reveals the controversies regarding the choice of its curation theme, architectural language, decision-making, and management models. Mechanisms such as individual preferences, disciplinary backgrounds, personal understanding of philosophical approaches to heritage, administrative structures, and religious faith exercise their causal powers on various site associations.

The complexities manifested in the actors' actions and effects demonstrate the ambiguous boundaries between the two dichotomies—the 'Eastern/Western' and the 'Tangible/Intangible'. This article has provided a vivid picture of how the perceived 'Western' and 'non-Western' approaches can be interwoven in individuals' actions, intentionally or unintentionally. It has also illustrated that tangible and intangible aspects of heritage are indeed inseparable, both in their interdependence and synchronicity. The controversies highlight that such initiatives must consider beyond the straitjacket of these dichotomies and adopt a more holistic and grounded approach to transforming these heritage sites towards a more sustainable future.

As the Project is still perceived to be a positive one by the administration, the issues regarding the project's impact can become more widespread if not given critical consideration. This article argues that future projects to initiate new associations must identify the actors forming these associations and what kind of causal powers are needed to sustain these associations. This insight will have broader relevance beyond this case study's context, scale, and heritage typology and potentially inform anticipatory policymaking in sustainability development and heritage management.

Funding: The WSA PhD Studentship and the Cardiff University Vice-Chancellor's International Scholarships for Research Excellence supported the research (application number for both fundings: 1752971).

Institutional Review Board Statement: The research has obtained ethics approval from the ethics committee at the Welsh School of Architecture, Cardiff University (approval number: EC1801.352).

Informed Consent Statement: All interviewees have given written consent, and their personal data are all anonymised in research outputs.

Data Availability Statement: Translated anonymised interview summaries, tables of identified actors and their opinions, and process diagrams for mapping the actors' voices and controversies can be accessed through this link: https://doi.org/10.6084/m9.figshare.25718766 (accessed on 29 April 2024). Further inquiries can be directed to the corresponding author.

Conflicts of Interest: The author declares no conflicts of interest.

Appendix A. Description of the Interventions of the Project

(number as indicated in Figure 9)

A small public square (1) is created in front of the entrance to the temple, next to the site of the dry Dragon Spring pond (2). The pond (Figure A1) was not refilled with water but only planted with grass which was said to be taken from the Yellow River's banks by the project team [73]. Three existing earthen caves (3), one of the common forms of vernacular architecture in the region, were dug into the earth mound where the temple is located are transformed into small exhibiting and resting spaces (Figure A2). The entrance to the temple was redesigned, different from both the entrances before the Project and from even earlier. The access from before the Project was an earth slope stretching from the east side of the pond up to the southeast corner of the temple ground.

Figure A1. The Dragon Spring Pond (*longquan chi*), March 2018 (source: author).

Figure A2. Renovated earthen caves by the entrance of Guangrenwang Temple, March 2018 (source: author).

According to the temple's caretaker interviewed during this research, at least as he remembered from the 1950s, the earlier entrance was a pathway across the middle of the Dragon Spring. It split into two ascending paths leading to the temple ground's southwest and southeast corners [81]. The new entrance takes visitors through a winding ascent (4) (Figure A3) to the ticket office, which is connected to a new office, a small community library and a living space for the caretakers (5). Visitors are then led to an enclosed introduction space (6). A full-scale section of the main hall is engraved on the ground, and a timeline on the wall shows where the construction date of the main hall is located in history relative to other well-known historic timber buildings in China. On the opposite wall is a plaque containing information about the Project, donors, and participants' names (Figure A4).

Figure A3. The winding ascent to the ticket office, March 2018 (source: author).

From the entrance to this point of the introduction space, the temple's main hall is blocked from the visitors' sight by the enclosing walls. Only a few occasional glimpses of the roof are possible, and only if one pays attention (Figure A5). This approach contrasts with the previous entrances, which led the visitors to the main temple ground via the side of the stage. The visitors are then directed to exit the introduction space and turn into a narrow ascending corridor (7). The walls that flank both sides of the corridor block out the other area of the temple and create a restricted frame pointing towards the side façade of the main hall, creating an image of the historic building's elevation.

While walking through the corridor, visitors are directed towards the next exhibition space (8) through a wall opening. The visitors are encouraged to turn towards this space after the corridor, where enlarged models of bracket sets of the surviving Tang structures in the country are displayed (Figure A6), with information and architectural drawings of these buildings exhibited on some permanently installed panels. This second exhibition space is connected to another long corridor (9) at the back of the temple. It takes the visitors to a raised platform (10) on the central axis of the temple, where an information panel explains the ruins of Wei City and Zhongtiao Mountain as a significant element of the historic setting for both the ancient city and the historic temple (Figure A7). The visitors are then encouraged to continue along the edge of the temple ground towards yet another corridor on the west side (11), where some brief information about other early timber structures in the region is displayed. Next to this corridor is a resting space (12). The museum space is separated from the main temple ground (13) by walls, except for a few openings. The circulation is guided along the site's edge rather than towards the centre, where the historic buildings stand. The historic steles are relocated and embedded into the wall behind the main hall (14). New statues of the deities in the main hall were also commissioned during the Project by Vanke [67].

Figure A4. Plaque with names of donors and participants of the Project, March 2018 (source: author).

Figure A5. Limited view of the main hall's roof from the entrance staircase, March 2018 (source: author).

Figure A6. View of the *dougong* model display through the wall opening of the corridor (source: author).

Figure A7. Information panel about the natural and historical setting of Guangrenwang Temple (information about the Ancient Wei City and Zhongtiao Mountain) with Zhongtiao Mountain in view (source: author).

Appendix B. Descriptions of the Surviving Historic Steles in the Temple

Four historic steles survive in the temple, two from the Tang Dynasty and two from the Qing Dynasty. The one from 808 CE records the water system in the area, which is the earliest record of the Dragon Spring (*longquan*) and the temple's initial construction. The 808 CE inscription is titled '*Guangrenwang Longquan Ji*', suggesting that Guangrenwang was already the subject of worship at the time. It also records that the temple was built next to the spring because it was believed to be where the deity resides, who can prevent draught. Another stele from 832 CE mentions the ancient Wei City and that there was a spring in the northwest corner of the city, which can be confirmed by archaeology (Figure 3). It records that the temple built in 808 resulted from drought and was commissioned by the former governor and that the temple buildings were already dilapidated. Another drought season between 831–832 CE prompted the temple's reconstruction by the villagers commissioned by Ruicheng's governor. The 832 inscription records that rain came pouring down upon this reconstruction and exclaimed the significance of paying tribute to nature. The surviving main hall is believed to be the result of this reconstruction. Both ninth-century steles mention that mountains surrounded the temple, and the area was crisscrossed with creeks. Indeed, Zhonglongquan (Middle Dragon Spring) village is only one of the three villages named after the Dragon Spring.[22] These historical records confirmed the close relationship between the temple, the local community, and the natural environment.

The two Qing Dynasty steles record two restorations of the temple. The 1758 one describes that the temple was restored in 1745, confirmed by the inscription inside the main hall. The stele describes that the temple was still well maintained, but the stage (*yuelou*) needed restoration by 1758. The village chiefs commissioned the repair of the stage and the east wall of the main hall, funded by several surrounding communities. The inscription also mentions that five communes (*she*) discussed and agreed that anyone who stored branches in the temple should be fined, suggesting the scope of the historic community that managed the temple. Besides the stage and the main hall, the inscription records a repair of another building and two corner gates. While it is unclear where the building was, it provided information on the historic access to the temple, possibly through two corner gates on both sides of the stage. Another stele from 1812 records that the temple needed another repair by 1806. It mentions the restoration in 1745 but not the one in 1758. The inscription describes that the restoration of the stage started in 1806 and was completed by 1812. The buildings were re-decorated, and the temple walls were repaired in 1811. The

1812 inscription also mentions that the repair was partly funded by selling a few trees in the temple, suggesting that trees were not only planted to provide timber for the repair but also used as a commodity to support the temple's maintenance. Another restoration in 1906 is recorded by an inscription written on a board underneath the ridge purlin of the main hall. A Qing Dynasty local chronography of Ruicheng mentions the Dragon Spring and records that it was connected to the Yellow River, which runs south of Ruicheng. The same entry mentions the temple as *Wulong Ci* (Wulong Temple). A village named *Houlongquan* village was also recorded in the chronography [120].

Notes

1 Due to the limited scope of this article, the theoretical underpinnings of the framework will not be elaborated on in detail. For further details of this framework see [50].

2 This department was renamed National Cultural Heritage Administration (NCHA) in English in 2018. However, SACH has been widely used in Anglophone academic literature. Therefore, SACH will still be used in this article to avoid confusion.

3 The re-discovery of Guangrenwang Temple is believed to be the result of the relocation of Yongle Temple from 1958–1964. The relocation was a major heritage project considered to be one of the most significant achievements at the beginning of the PRC. Guangrenwang Temple, being very close to the new site of Yongle Temple, was 're-discovered' by those who participated in the relocation project in 1958.

4 With many existing contested opinions, the most common understanding is that there are about four and a half Tang timber structures left in the country, which are the main hall of Nanchan Temple, the east main hall of Foguang Temple, the main hall of Tiantai Monastery, the main hall of Guangrenwang Temple (all in Shanxi Province), and the ground floor of the bell tower of Kaiyuan Temple in Hebei Province. While the first two are recognised as Tang structures without much dispute (although the first one had been significantly restored in 1978), there is no definite evidence for the construction dates of the surviving main halls of Tiantai Monastery and Guangrenwang Temple. During the Southern Project, new evidence was found in Tiantai Monastery which suggested that its main hall was constructed during the Five Dynasties instead. It makes it more desirable for the state and population that the main hall of Guangrenwang Temple retains its Tang structure status.

5 On one of the beams inside the main hall, an inscription reads "*Yi Jiu Wu Ba Nian Shi Yi Yue Shi Jiu Ri Shan Xi Sheng Wen Wu Guan Li Wei Yuan Hui Rui Cheng Xian Ren Min Wei Yuan Hui Chong Xiu Guang Ren Miao Ji Nian*" (To commemorate the restoration of Guangren Temple on November 19th 1958 by Shanxi Cultural Relics Management Committee and Ruicheng County People's Committee).

6 This interpretation, however, is not the original conception of Libeskind when he designed the building. Interestingly, the 'reinterpretation' of the building as a dragon by *Vanke* and its subsequent link with the crowd-funding scheme of the Guangrenwang (Five-Dragon) Temple adds another 'make-believe' aspect to the Project.

7 Wang, Lu, Dou, and Zhou, all of whom are from an architectural background, consider the Project an excellent opportunity to transform and enhance the site's status as a place for knowledge transfer, highlighting its significance in architectural history. Specifically, Zhou considers that instead of revitalising the temple as a religious space for worshipping deities, the project managed to recreate a secular sacred space by presenting knowledge as the subject of worship in a museum setting [88]. This perspective is resonated by Lu, who considers that knowledge has replaced religion as a driving factor for the meaning-making process on-site [89]. Lv (Lyu), coming from the same premise, takes a more cautious stance as a heritage professional. He admits that since there is not enough research on the temple's religious history, folk customs and rituals, emphasising its significance in architectural history is a reasonable choice [84].

8 This concern is raised by Liu Diyu, who acknowledges the subjectivity of value assessment. He warns that overly emphasising the authorised value assessment that is commonly known to the general public may exclude the possibility to discover lesser-known and hidden meanings [86]. A similar comment was made by Zhang Lufeng, who considers that the existence of a heritage site is a composition of various meanings which should allow multiple interpretations and the new intervention should be more open-ended. He questions the decision to transform the historic temple to a museum, which highlights the site's 'contemporary values' but might have excluded others [86].

9 While this value assessment should be publicly available, upon consulting the inventory of the national PCHS, it is not obvious where the said value assessment is.

10 Interestingly, there are a few commentators who acknowledge the significance of the temple's religious connotation among the current local population but consider that the Project has indeed managed to elevate this discourse. By referencing Article 7 and Article 33 of the Nairobi Declaration, Guo, as a heritage professional, acknowledges that Guangrenwang Temple is both a space of worship and a place for knowledge. He comments that the Project, while revitalising the heritage site, also revived the faith for local religions and culture among the local population. He also suggests that the local community is not bothered by the new layers added to the identities of the temple. However, the article has not provided any evidence in support of these statements [85].

11   A candid comment from the village chief reveals the complexity behind this silence. While recalling his experience of being invited to Milan as the representative of the villagers to speak at Vanke's exhibition at the Expo, he admitted that he was a little nervous because it involved discussion of issues relating to religion and faith. He exemplified that a local official was deposed simply because he gave a speech at a temple fair [67]. The village chief's comment is emblematic of the deliberate ambiguity in the implementation of religious policy across the country.

12   These opinions resemble those of some community and public members, as well as some heritage professionals. The leading engineer of the Southern Project quoted her colleague's exclaims upon seeing the site during the interview of this research—'(It is) so twee! So full of petit-bourgeois sentiments! (in Chinese)'—while commenting that it is telling that these adjectives which are usually used for describing urban lifestyles were inspired by this little village temple [75]. This opinion is echoed by Liu Diyu, who comments that the architectural language of the museum, including the scale, the volume, and the materials of the floor tiles on the temple ground, resembles the design of an urban square or a park [86].

13   Lv points out that getting private funding and participation is beneficial to tackling the shortage of state funding for the caretaking of heritage sites, especially on projects that do not involve the historic buildings. More importantly, he considers that getting more actors from society to participate is itself a process which enriches the social value of the heritage site. In the same way, the fact that the Project has attracted the attention of the general public has a similar effect [84]. This perspective has its root in a significant shift in China's heritage discourse. The addition of the categories of social and cultural values in the China Principles is an attempt to incorporate intangible associations of these tangible sites into their value assessment process.

14   An observation by a Weibo user Chinn-秦汇川's visit to the temple in 2019 provides a vivid account and insights into the reasons behind. According to their report, when asked why they do not go to the museum even though they approve of the positive impact on the temple's environment, the villagers replied, '*Sure, it is beautiful. But it is not so interesting for us.*''*We do not understand it anyway.*''*It is hard to find people who are under 60 years old in the village. We cannot climb those stairs.*''*It is hardly as lively as here (around 60 metres from the temple).*'

15   However, they suggested that the villagers could have been invited to participate in the construction of the walls, which still did not address the most fundamental aspect of the issue, the lack of community involvement in the decision-making of the Project.

16   The term 'dead heritage' is referring to a common situation in many sites' post-restoration status in the Southern Project, where they are closed and are referred to as 'museum objects' locked away from the public [86,88].

17   This decrease of funding is a nation-wide policy which is meant to encourage the provincial and local governments to take up more responsibilities in the management of national PCHS. It is part of the administration's 'decentralisation process'.

18   Meanwhile, there is much debate on whether the effort of revitalisation could be shared by different sectors of society besides the public sector. As mentioned in Section 3.3.4, some heritage professionals believe the participation of the private sector in the case of Guangrenwang Temple has increased the social impact of the heritage site. They believe that the broader involvement of society in heritage projects is beneficial for their long-term survival. During the interviews of this research, most of the local officials tend to think that there should be a certain degree of control over these projects by the administrations since these sites are 'very important' [81,93,94,114]. This opinion also represents a general attitude towards private sector participation in heritage projects among heritage professionals, caretakers, and community members. What differs between individual interviewees is the extent to which the government should be in control. Some consider it only suitable for the private sector to get involved financially, while others consider it essential for the private sector to come up with viable management and operational plans and that they should oversee the implementation and sustaining the management of the site in the long term.

19   As suggested in national legislation and the China Principles, non-profit functions such as research institutes, museums, and community centres are preferred as 'appropriate use' of heritage sites [71,87]. It is reasonable to question the viability and sustainability of funding and human resources for these entities.

20   The agreement between Vanke and the public sector is that while Vanke was responsible for financing and implementing the Project, it is not taking on the responsibility of running the museum, which would be given back to the local authority instead. From what can be seen on-site, such a model does not guarantee the continuous innovation and maintenance of the museum. As soon as the Project was completed, the management model went back to being almost the same as it was previously. According to the local official and the caretaker, events and activities only happen sporadically on-site [81,82].

21   According to one unnamed local official, such a project is like 'gifting a low-income family a big refrigerator. Even though it might seem like a nice gesture, the low-income family now has to carry the burden of buying more food to put in it and paying for the electricity bill' [79]. Although the local official allegedly said so because he was 'not understanding what the project was actually about', such an analogy rings true considering the reality seen on the ground during this research [67].

22   The other two villages are Qianlongquan (front Dragon Spring) village and Houlongquan (back Dragon Spring) village.

References

1.   Yan, H. *World Heritage Craze in China—Universal Discourse, National Culture, and Local Memory*; Berghahn Books: New York, NY, USA, 2018.

2.   Zhu, Y.; Maags, C. *Heritage Politics in China: The power of the Past*; Routledge: London, UK, 2020.

3.   Nara Conference on Authenticity. In *The Nara Document on Authenticity*; UNESCO: Nara, Janpan, 1994.

4. Akagawa, N. Rethinking the global heritage discourse—Overcoming 'East' and 'West'? *Int. J. Herit. Stud.* **2016**, *22*, 14–25. [CrossRef]
5. Falser, M.S. From Venice 1964 to Nara 1994—Changing concepts of authenticity? In *Conservation and Preservation: Interactions between theory and practice: In memoriam Alois Riegl, Proceedings of the International Conference of the ICOMOS International Scientific Committee for the Theory and the Philosophy of Conservation and Restoration, 23–27 April 2008, Vienna, Austria, 2010*; Falser, M.S., Lipp, W., Tomaszewski, A., Eds.; ICOMOS: Veinna, Austria, 2010.
6. Lowenthal, D. *The Past Is a Foreign Country*; Cambridge University Press: Cambridge, UK, 1985.
7. Ryckmans, P. *The Chinese Attitude towards the Past*; The Australian National University: Canberra, Australia, 1986.
8. Smith, L. *Archaeological Theory and the Politics of Cultural Heritage*; Routledge: London, UK, 2004.
9. Waterton, E.; Smith, L. The recognition and misrecognition of community heritage. *Int. J. Herit. Stud.* **2010**, *16*, 4–15. [CrossRef]
10. Zhu, Y. Cultural effects of authenticity: Contested heritage practices in China. *Int. J. Herit. Stud.* **2015**, *21*, 594–608. [CrossRef]
11. Gentry, K.; Smith, L. Critical heritage studies and the legacies of the late-twentieth century heritage canon. *Int. J. Herit. Stud.* **2019**, *25*, 1148–1168. [CrossRef]
12. Matsuda, A.; Mengoni, L.E. Introduction: Reconsidering cultural heritage in East Asia. In *Reconsidering Cultural Heritage in East Asia*; Matsuda, A., Mengoni, L.E., Eds.; Ubiquity Press: London, UK, 2016.
13. Gao, Q.; Jones, S. Authenticity and heritage conservation: Seeking common complexities beyond the 'Eastern' and 'Western' dichotomy. *Int. J. Herit. Stud.* **2020**, *27*, 90–106. [CrossRef]
14. Sand, J. Japan's Monument Problem: Ise Shrine as Metaphor. *Past Present* **2015**, *226* (Suppl. S10), 126–152. [CrossRef]
15. Stubbs, J.H.; Thomson, R.G. *Architectural Conservation in Asia: National Experiences and Practice*; Routledge: London, UK, 2017.
16. Winter, T. Beyond Eurocentrism? Heritage conservation and the politics of difference. *Int. J. Herit. Stud.* **2014**, *20*, 123–137. [CrossRef]
17. Smith, L. *Uses of Heritage*; Routledge: Oxon, UK, 2006.
18. Forster, A.M.; Thomson, D.; Richards, K.; Pilcher, N.; Vettese, S. Western and Eastern Building Conservation Philosophies: Perspectives on Permanence and Impermanence. *Int. J. Archit. Herit.* **2018**, *13*, 870–885. [CrossRef]
19. Dirlik, A. Chinese History and the Question of Orientalism. *Hist. Theory* **1996**, *35*, 96–118. [CrossRef]
20. Sen, A. *Development as Freedom*; Oxford University Press: Oxford, UK; New York, NY, USA, 1999.
21. Lowenthal, D. *The Heritage Crusade and the Spoils of History*; Cambridge University Press: Cambridge, UK, 1998.
22. Kirshenblatt-Gimblett, B. From ethnology to heritage: The role of the museum. *SIEF Keynote Marseilles*, April 2004, pp. 1–8. Available online: https://scholar.archive.org/work/bvomhvpyjvgehjpnbxo2kvsqk4/access/wayback/http://www.nyu.edu/classes/bkg/web/SIEF.pdf (accessed on 29 April 2024).
23. Rudolff, B. Intangible'and 'Tangible'Heritage: A Topology of Culture in Contexts of Faith. Ph.D. Thesis, Johannes Gutenberg-University of Mainz, Mainz, Germany, 2006.
24. Ruggles, D.F.; Silverman, H. *Intangible Heritage Embodied*; Springer: Berlin/Heidelberg, Germany, 2009.
25. Smith, L.; Akagawa, N. *Intangible Heritage*; Routledge: London, UK, 2009.
26. State Administration of Cultural Heritage. *China's Cultural Heritage Enterprise in the 30 Years of 'Reform and Opening Up'*; State Administration of Cultural Heritage: Beijing, China, 2008.
27. Gruber, S. Protecting China's Cultural Heritage Sites in Times of Rapid Change: Current Developments, Practice and Law. *Asia Pac. J. Environ. Law* **2007**, *10*, 253–301.
28. Maags, C. Disseminating the policy narrative of 'Heritage under threat' in China. *Int. J. Cult. Policy* **2020**, *26*, 273–290. [CrossRef]
29. Harrison, R. *Heritage: Critical Approaches*; Routledge: Oxon, UK, 2013.
30. Smith, S.A. Contentious Heritage: The Preservation of Churches and Temples in Communist and Post-Communist Russia and China. *Past Present* **2015**, *226*, 178–213. [CrossRef]
31. State Administration of Cultural Heritage. *Information Database of National PCHS*; State Administration of Cultural Heritage: Beijing, China, 2020.
32. Zhu, Y.; Li, N. Groping for stones to cross the river: Governing heritage in Emei. In *Cultural Heritage Politics in China*; Springer: Berlin/Heidelberg, Germany, 2013; pp. 51–71.
33. Zhao, W. Local versus national interests in the promotion and management of a heritage site: A case study from Zhejiang Province, China. In *Cultural Heritage Politics in China*; Springer: Berlin/Heidelberg, Germany, 2013; pp. 73–100.
34. Yan, H. World Heritage as discourse: Knowledge, discipline and dissonance in Fujian Tulou sites. *Int. J. Herit. Stud.* **2015**, *21*, 65–80. [CrossRef]
35. Maags, C. Creating a Race to the Top: Hierarchies and Competition within the Chinese ICH Transmitters System. In *Chinese Cultural Heritage in the Making: Experiences, Negotiations and Contestations*; Maags, C., Svensson, M., Eds.; Amsterdam University Press: Amsterdam, The Netherlands, 2018; pp. 121–144.
36. Lai, G. The emergence of 'cultural heritage' in modern China: A historical and legal perspective. In *Reconsidering Cultural Heritage in East Asia*; Ubiquity Press: London, UK, 2016.
37. Zhu, Y. Thoughts on the Comparison of Heritage Conservation in the West and the East: A Cross-cultural Perspective. *Southeast Cult. (Dongnan Wenhua)* **2011**, 118–122.

38. Zhu, Y. Authenticity and Heritage Conservation in China: Translation, Interpretation, Practices. In *Authenticity in Architectural Heritage Conservation Discourses, Opinions, Experiences in Europe, South and East Asia*; Weiler, K., Gutschow, N., Eds.; Springer: Berlin/Heidelberg, Germany, 2017; pp. 187–200.

39. D'Ayala, D.; Wang, H. Conservation Practice of Chinese Timber Structures. *J. Archit. Conserv.* **2006**, *12*, 7–26. [CrossRef]

40. Zhang, S. The Development and Institutional Characteristics of China's Built Heritage Conservation Legislation. *Built Herit.* **2022**, *6*, 11. [CrossRef]

41. Chen, W. The Study on the Conservation Theory and Method of Heritage. Ph.D. Thesis, Chongqing University, Chongqing, China, 2006.

42. Xu, Y. The Temples of the Five Dynasties, the Song Dynasty and the Jin Dynasty in Changzhi and Jincheng District. Ph.D. Thesis, Peking University, Beijing, China, 2003.

43. DeLanda, M. *A New Philosophy of Society: Assemblage Theory and Social Complexity*; Bloomsbury Publishing: London, UK, 2019.

44. Archer, M.; Bhaskar, R.; Collier, A.; Lawson, T.; Norrie, A. *Critical Realism: Essential Readings*; Routledge: London, UK, 2013.

45. Bhaskar, R. *A Realist Theory of Science*; Routledge: London, UK, 2008. [CrossRef]

46. Bhaskar, R. *Enlightened Common Sense: The Philosophy of Critical Realism*; Routledge: London, UK, 2016.

47. Bhaskar, R. *The Possibility of Naturalism—A Philosophical Critique of the Contemporary Human Sciences*, 4th ed.; Routledge: Oxon, UK, 2015.

48. Bhaskar, R.; Danermark, B.; Price, L. *Interdisciplinarity and Wellbeing: A Critical Realist General Theory of Interdisciplinarity*; Taylor & Francis: Abingdon, UK, 2017.

49. Tam, L. Sustainable Heritage Management in Contemporary China. Ph.D. Thesis, Cardiff University, Cardiff, UK, 2022.

50. Byrne, D.; Harvey, D.L.; Mjoset, L.; Carter, B.; Sealey, A.; Williams, M.; Dyer, W.; Elman, C.; Uprichard, E.; Phillips, D.; et al. *The SAGE Handbook of Case-Based Methods*; SAGE Publications: London, UK, 2009.

51. Elder-Vass, D. *The Causal Power of Social Structures: Emergence, Structure and Agency*; Cambridge University Press: Cambridge, UK, 2010.

52. Elder-Vass, D. Searching for realism, structure and agency in Actor Network Theory. *Br. J. Sociol.* **2008**, *59*, 455–473. [CrossRef] [PubMed]

53. Danermark, B.; Ekström, M.; Karlsson, J.C. *Explaining Society: Critical Realism in the Social Sciences*; Routledge: London, UK, 2002.

54. Yin, R.K. *Case Study Research: Design and Methods*; Sage: London, UK, 2008.

55. Sandelowski, M. Focus on research methods: Whatever happened to qualitative research. *Res. Nurs. Health* **2000**, *23*, 334–340. [CrossRef]

56. Yaneva, A. *Mapping Controversies in Architecture*; Routledge: London, UK, 2016.

57. Chai, Z. Discussion on the 'not changing the original form of cultural heritage' in the restoration of historic buildings. In *Collected Works of Chai Zejun on Ancient Architecture*; Cultural Relics Publishing House: Beijing, China, 1999; pp. 303–324.

58. Jiu, G. Wulong Temple to the south of Zhongtiao Mountain in Shanxi. *Cult. Relics* **1959**, 43–44. (In Chinese)

59. Chai, Z. A brief documentation of the survey and conservation of Shanxi's ancient architecture and its affiliated cultural relics in the past 30 years. In *Collected Works of Chai Zejun on Ancient Architecture*; Zhou, C., Ed.; Cultural Relics Publishing House: Beijing, China, 1999; pp. 10–31.

60. GADM. *China GADM Data*; GADM, 2021. Available online: https://gadm.org/ (accessed on 5 June 2021).

61. Google Maps. Satellite map of Zhonglongquan Village, Ruicheng. Data from CNES/Airbus & Maxar Technologies, 2021.

62. Li, Z. Research of the Tang Dynasty Wulong Temple and its Music Tower Stele in Ruicheng, Shanxi. *China Tradit. Opera* **2015**, *2*, 92–103. [CrossRef]

63. Qi, X. It should have been a spiritual conversation—Thoughts on Wulong Temple's environment improvement. In *Chinese National Geography Bishan 12—Architects in the Countryside*; CITIC Press Group: Beijing, China, 2020; pp. 102–111.

64. Chai, Z. Several Significant Ancient Architecture Examples in Shanxi. In *Collected Works of Chai Zejun on Ancient Architecture*; Zhou, C., Ed.; Cultural Relics Publishing House: Beijing, China, 1999; pp. 148–199.

65. Fu, X. *Chinese Ancient Architectural History—Architecture of the Two Jin Dynasties, Northern and Southern Dynasties, Sui Dynasty, Tang Dynasty, and the Five Dynasties*; China Architecture and Building Press: Beijing, China, 2001; Volume 2.

66. He, D. The Tang Dynasty Timber Main Hall in Guangrenwang Temple in Ruicheng, Shanxi. *Cult. Relics* **2014**, 69–80.

67. Jia, D. Ecology: The rebirth of a Tang Dynasty temple: Cultural relics and the local village life. In *Sanlian Life Weekly*; SDX Joint Publishing: Beijing, China, 2017; pp. 86–95.

68. State Administration of Religious Affairs. Registered Religious Venues in Ruicheng County Database. 2020. Available online: https://www.sara.gov.cn/resource/common/zjjcxxcxxt/zjhdcsjbxx.html (accessed on 5 June 2021).

69. Liang, C. Guangrenwang Temple in Ruicheng Experiments Crowd-Funding for Heritage Conservation. *Shanxi Evening News*. 29 July 2015. Available online: https://xw.qq.com/cmsid/20150729038137/undefined (accessed on 9 April 2019).

70. National People's Committee of PRC. *Cultural Heritage Protection Law of the People's Republic of China (2017 Amendment)*; The NPC Standing Committee: Beijing, China, 2017.

71. Hu, J. The Oldest Surviving Taoist Temple—Shanxi Guangrenwang Temple Opens Its Gate to Guests after Restoration. *China News*, 4 January 2015. Available online: https://www.chinanews.com.cn/cul/2015/01-04/6933778.shtml (accessed on 5 April 2019).

72. Guo, G. The Environment Upgrading Project of Guangrenwang Temple in Ruicheng, Shanxi Is Completed. *China Cultural Relics News*, 20 May 2016.
73. Wang, H. Theoretic Propositions on the Design of the Five-Dragons Temple Environmental Improvement. *World Archit.* **2016**, *7*, 110–114+127. [CrossRef]
74. Heritage professional 5. Interview with heritage professional 5. Tam, L. 2018. Available online: https://figshare.com/articles/dataset/Data_for_the_article_A_Controversial_Make-over_of_a_Make-believe_heritage_The_Transformation_of_Guangrenwang_Temple/25718766 (accessed on 29 April 2024).
75. UED Magazine. The Warmest Project of 2016!—The 'Rammed Earth' Story of Wulong Temple's Environment Improvement. *UEDMegazine Wechat*, 26 June 2016.
76. Archdaily. Five-Dragons Temple Environmental Refurbishment/URBANUS. *Archdaily*, 2 May 2017. Available online: https://www.archdaily.com/870312/the-design-of-the-five-dragons-temple-environmental-improvement-urbanus#:~:text=The%20entrance%20of%20the%20Five,the%20side%20of%20the%20path. (accessed on 29 April 2024).
77. China News. For a Temple, Vanke Devoted a Lot. *Chinanews.com*, 27 September 2015.
78. AC Editorial. The Country and Elites in Front of Wulong Temple. *ArchiCreation*, 17 May 2016.
79. Heritage Professional 3. Interview with Heritage Professional 3. Tam, L., 2018. Available online: https://figshare.com/articles/dataset/Data_for_the_article_A_Controversial_Make-over_of_a_Make-believe_heritage_The_Transformation_of_Guangrenwang_Temple/25718766 (accessed on 29 April 2024).
80. Local Official 9. Interview with Local Official 9. Tam, L., 2018. Available online: https://figshare.com/articles/dataset/Data_for_the_article_A_Controversial_Make-over_of_a_Make-believe_heritage_The_Transformation_of_Guangrenwang_Temple/25718766 (accessed on 29 April 2024).
81. Caretaker of YC Temple 4. Interview with Caretaker of YC Temple 4. Tam, L., 2018. Available online: https://figshare.com/articles/dataset/Data_for_the_article_A_Controversial_Make-over_of_a_Make-believe_heritage_The_Transformation_of_Guangrenwang_Temple/25718766 (accessed on 29 April 2024).
82. Huang, J. Architectural Review. *World Archit.* **2016**, *7*, 117.
83. Lyu, Z. Discussions on the Five-Dragons Temple Environmental Improvement Project. *World Archit.* **2016**, *7*, 116.
84. Guo, L. Value, Interpretation and Authenticity: Reflection on the Event of Environmental Improvement of Five-Dragon Temple. *World Archit.* **2017**, *8*, 124–127. [CrossRef]
85. The "death" and "life" of architectural heritage—The symposium on the environmental enhancement of the Temple of Five Dragons. *Archit. J.* **2016**, *8*, 1–7.
86. ICOMOS China. Principles for the Conservation of Heritage Sites in China. 2015. Available online: http://hdl.handle.net/10020/gci_pubs/china_principles_2015 (accessed on 29 April 2024).
87. Zhou, R. Dragon as a vibrant element—The critical replay of the environmental enhancement of the Temple of Five Dragons. *Archit. J.* **2016**, *8*, 14–17.
88. Lu, A. Archaeological Architecture and Artificial Situation—Thoughts on the Renovation Design of Five Dragons Temple. *Time+Archit.* **2016**, *4*, 123–130. [CrossRef]
89. Dou, P. Temple or Knowledge? *World Archit.* **2016**, *7*, 117.
90. State Administration of Cultural Heritage. *Official Reply to the Environment Improvement Project of Guangrenwang Temple*; Department of Heritage Conservation and Archaeology: Beijing, China, 2015.
91. State Administration of Cultural Heritage. *Opinions on the Design Proposal of the Environment Improvement Project of Guangrenwang Temple*; Department of Heritage Conservation and Archaeology: Beijing, China, 2015.
92. Local Officials 1&2. Interview with Local Officials 1&2. Tam, L., 2018. Available online: https://figshare.com/articles/dataset/Data_for_the_article_A_Controversial_Make-over_of_a_Make-believe_heritage_The_Transformation_of_Guangrenwang_Temple/25718766 (accessed on 29 April 2024).
93. Local Official 10. Interview with Local Official 10. Tam, L., 2018. Available online: https://figshare.com/articles/dataset/Data_for_the_article_A_Controversial_Make-over_of_a_Make-believe_heritage_The_Transformation_of_Guangrenwang_Temple/25718766 (accessed on 29 April 2024).
94. Koepnick, L. Forget Berlin. *Ger. Q.* **2001**, *74*, 343–354. [CrossRef]
95. Hou, Z. Long·Plan: The Destined Relationship between the Five-Dragons Temple and the Vanke Pavilion. *World Archit.* **2016**, *7*, 115. [CrossRef]
96. Li, X. Social participation: A necessary force for cultural heritage conservation and utilisation. *China Cultural Relics News*, 17 May 2019.
97. Liu, A.; Yu, B. Practices and reflections on the society's participation in cultural heritage conservation and utilisation. *China Anc. City* **2018**, 77–82+58.
98. Yu, J. Reflections on the feasibility of introducing forces of the society into cultural heritage conservation. *World Antiq.* **2018**, *4*, 56–58.
99. Zhang, B. The society is the living force to inherit cultural heritage conservation. *People's Congr. China* **2017**, *24*, 32–33.

100. Liu, A. The development trend of the society's participation in cultural heritage conservation—A few case studies. In Proceedings of the Innovation and Development of Social Organisation for Cultural Heritage Conservation—The Second Symposium on Social Participation in Cultural Heritage Conservation and Utilisation, Shanghai, China, 11 August 2017; China Foundation for Cultural Heritage Conservation, Ed.; Cultural Relics Publishing House: Shanghai, China, 2017; pp. 53–61.
101. Shen, X. A Study on Social Participation in Cultural Heritage Protection. *J. Zhejiang Int. Stud. Univ.* **2017**, *6*, 103–109.
102. Feng, Y. Using the power of enterprises to implement the public participation of cultural heritage conservation. In Proceedings of the Forum of Public Participation in Cultural Heritage Conservation, 3 November 2016; China Foundation for Cultural Heritage Conservation, Ed.; Cultural Relics Publishing House: Beijing, China, 2016; pp. 159–163.
103. Yang, J. The difficulties and paths for the general public to participate in cultural heritage conservation. *Yindu J.* **2014**, *35*, 116–118. [CrossRef]
104. Sun, J. Cultural Heritage Conservation Calls for Society Participation. *China Culture Daily*, 18 December 2014; pp. 1–3.
105. Ning, L. The society should be encouraged to take part in the conservation of historic architecture. *China Cultural Daily*, 29 May 2014; pp. 1–2.
106. Qi, R. The Social Participation of Immovable Cultural Heritage Conservation. *China Cultural Relics News*, 25 January 2013; pp. 1–3.
107. Wang, Y. Co-own, co-protect, co-share—Reflections on public participation in cultural heritage conservation (II). *China Cult. Herit. Sci. Res.* **2010**, *3*, 12–20.
108. Li, Y. Practices and reflections on getting the society to participate in cultural heritage conservation. *China Cult. Herit. Sci. Res.* **2006**, *4*, 36–38+43.
109. Provincial Official 3. Interview with Provincial Official 3 in Shanxi. Tam, L., 2018. Available online: https://figshare.com/articles/dataset/Data_for_the_article_A_Controversial_Make-over_of_a_Make-believe_heritage_The_Transformation_of_Guangrenwang_Temple/25718766 (accessed on 29 April 2024).
110. Heritage Professional 2. Interview with Heritage Professional 2. Tam, L., 2018. Available online: https://figshare.com/articles/dataset/Data_for_the_article_A_Controversial_Make-over_of_a_Make-believe_heritage_The_Transformation_of_Guangrenwang_Temple/25718766 (accessed on 29 April 2024).
111. Heritage Professional 1. Interview with the Manager of the Information and Documentation Project of the Southern Project. Tam, L., 2018. Available online: https://figshare.com/articles/dataset/Data_for_the_article_A_Controversial_Make-over_of_a_Make-believe_heritage_The_Transformation_of_Guangrenwang_Temple/25718766 (accessed on 29 April 2024).
112. Provincial Official 2. Interview with Provincial Official 2. Tam, L., 2018. Available online: https://figshare.com/articles/dataset/Data_for_the_article_A_Controversial_Make-over_of_a_Make-believe_heritage_The_Transformation_of_Guangrenwang_Temple/25718766 (accessed on 29 April 2024).
113. Local Official 3. Interview in Changzhi Municipality. Tam, L., 2018. Available online: https://figshare.com/articles/dataset/Data_for_the_article_A_Controversial_Make-over_of_a_Make-believe_heritage_The_Transformation_of_Guangrenwang_Temple/25718766 (accessed on 29 April 2024).
114. Hampton, M.P. Heritage, local communities and economic development. *Ann. Tour. Res.* **2005**, *32*, 735–759. [CrossRef]
115. Salazar, N.B. The glocalisation of heritage through tourism: Balancing standardisation and differentiation. In *Heritage and Globalisation*; Labadi, S., Long, C., Eds.; Routledge: London, UK, 2010; pp. 130–147.
116. McGuigan, J. *Rethinking Cultural Policy*; McGraw-Hill Education: London, UK, 2004.
117. Cole, S.; Morgan, N. *Tourism and Inequality: Problems and Prospects*; CABI: Wallingford, UK, 2010.
118. Urry, J.; Larsen, J. *The Tourist Gaze 3.0*; Sage: Thousand Oaks, CA, USA, 2011.
119. Tam, L. Something More Than a Monument—The Long-term Sustainability of Rural Historic Temples in China. *Religions* **2019**, *10*, 289. [CrossRef]
120. Mo, P. *Chronography of Ruicheng County, Haizhou in the Qianlong Era*; Yan, R., Ed.; 1764.

Disclaimer/Publisher's Note: The statements, opinions and data contained in all publications are solely those of the individual author(s) and contributor(s) and not of MDPI and/or the editor(s). MDPI and/or the editor(s) disclaim responsibility for any injury to people or property resulting from any ideas, methods, instructions or products referred to in the content.

 *architecture*

Article

Sustainable Urban Heritage: Assessing Baghdad's Historic Centre of Old Rusafa

Mazin Al-Saffar 

Manchester School of Architecture, Manchester Metropolitan University, Manchester M1 7ED, UK; m.al-saffar@mmu.ac.uk

Abstract: Baghdad's historical centre is Old Rusafa, which has a long history dating back over a thousand years. The area enclosed within the old wall is approximately 5.4 square kilometres and contains nearly 15,700 buildings. The city's old core contains significant heritage buildings that belong to the Abbasid Empire (762–1258) and the Ottoman Period (1638–1917). This paper assesses Baghdad's historical centre and urban heritage. It addresses how the urban fabric has faced irreparable damage, a weak definition of demands, and an ambiguous formulation of what to preserve. The research examines Old Rusafa's dense irregular fabric, significant old souqs, heritage mosques, historical buildings, and traditional Baghdadi courtyard houses. The research implements various research strategies at different levels to evaluate the current condition of the built heritage in the city centre. It adopts a mixed methodological research approach that brings information from both qualitative and quantitative methods to address the research problems. The paper argues that achieving sustainable urban heritage requires considering efficient and sustainable strategies that drive urban evolution and encourage historic centre revitalisation towards sustainable heritage conservation. The outcomes of this paper raise awareness of the significance of safeguarding Baghdad's Islamic architecture and the sustainable reuse of its uniquely built heritage stock.

Keywords: sustainable heritage conservation; urban heritage; architecture heritage; city centre; historic city; sustainable future; conservation and participation; critical heritage theory; cultural heritage and intangible heritage; tangible and intangible heritage

Citation: Al-Saffar, M. Sustainable Urban Heritage: Assessing Baghdad's Historic Centre of Old Rusafa. *Architecture* **2024**, *4*, 571–593. https://doi.org/10.3390/architecture4030030

Academic Editor: Avi Friedman

Received: 26 April 2024
Revised: 23 July 2024
Accepted: 7 August 2024
Published: 9 August 2024

Copyright: © 2024 by the author. Licensee MDPI, Basel, Switzerland. This article is an open access article distributed under the terms and conditions of the Creative Commons Attribution (CC BY) license (https://creativecommons.org/licenses/by/4.0/).

1. Introduction

Baghdad is one of the leading cultural centres in the Middle East. It has been a centre of political and economic operations since it was chosen by Caliph Al-Mansur as his capital city for the Abbasid Empire in 762 CE. Up to the 21st century, the city had been occupied multiple times by different groups, such as Mongol leader Hülegü in 1258, the Ottomans (1638–1917), the British (1917–1932), and the Americans (2003), who have all left their marks in varying degrees. Baghdad's social structure is complex and has different types of communities that construct the city's cultural identity. During the 20th and 21st centuries, civil wars, migration out of the city, and an unstable political climate heavily affected Baghdad's cultural heritage. Since the 1950s, the internal and external immigration of original communities out of Baghdad's city centre—such as Muslims, Christians, and Jews—has been one of the issues that impacted the social demographic change in the city, where they left behind many heritage buildings and houses. Some of these heritage assets are still empty and have ultimately deteriorated or face the threat of demolition.

Over the last five decades, many studies and projects have been carried out by UNESCO, ICOMOS, and other organisations to conserve historic places in the world's cities. Iraq has participated in the efforts of UNESCO for the preservation of its cultural heritage by proposing several sites for inscription to the World Heritage Committee as having outstanding universal value and some of the most important archaeological collections in the world (Hatra, Ashur, Samarra, traditional Iraqi houses, and the Iraqi Museum) [1]. A total of 20,000 sites were estimated by Iraqi heritage experts to require protection due to the risk of significant deterioration. So far, about 700 archaeological sites have been discovered in Baghdad alone [2,3]. Therefore, the preservation of Iraq's and Baghdad's heritage is quite complex and faces multiple challenges. This paper addresses some of these challenges and assesses how they impact the built heritage of Baghdad. In addition, the paper also explores the relationship between urban heritage and socio-economic constraints. It argues that a better understanding of urban heritage will depend upon understanding both its heritage context, the categories of heritage values, and influences from more contextualised socio-economic and environmental forces.

1.1. Materials and Methods

This research is based on a combination of theoretical and empirical data to ensure that research questions are answered by using appropriate methodologies. A mixed methodology is applied, which combines quantitative and qualitative processes of analysis to assess the old centre of Baghdad's current situation, problems, and challenges [4,5]. In addition, a case study method is employed as a research tool to analyse the Old Rusafa built heritage [6].

Specifically, this paper focusses on the physical built heritage in an area located between Al-Rashid Street and the Tigris Riverfront in Old Rusafa. In this research, various assessment strategies are implemented, including a walking method, a serial vision method, and an observation ethnographic approach—all of which assess the built heritage situated within the old centre [7]. Qualitative and quantitative approaches are ultimately two different methodological systems selected to accomplish the aim of the research study [8]. The nature of this study can be defined as a multi-strategy approach, where each method integrates and builds on the strength of the other [9].

In this paper, the information has been collected from diverse sources associated with the various departments of the Municipalities of Baghdad, the Presidency of the Mayoralty of Baghdad, the Baghdad Heritage Department, the Urban Planning Department of Bagdad, the University of Technology, and the University of Baghdad. In all phases of the case, a wide variety of data from different sources has been integrated; however, the source and type of data depend on the case and its nature [4]. Therefore, investigating the historic centre of Old Rusafa (the area between Al-Rashid Street and the Riverfront) as a case study area can produce a comprehensive understanding of the particular needs for sustainable urban heritage development in one of the significant historical zones (Figure 1).

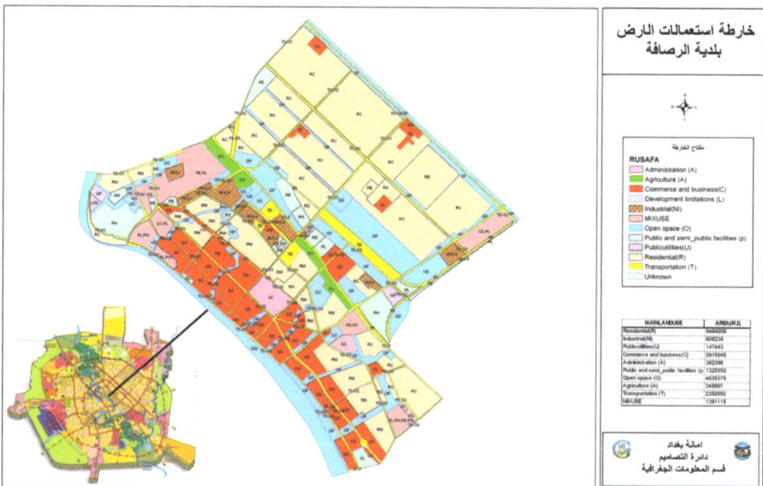

Figure 1. Rusafa District—Old Rusafa land use. Source: author, according to [10].

1.2. Conservation and Cultural Built Heritage

Substantial changes in the urbanised world have led to the formation and necessity of architectural and urban conservation. Many challenges emerge in the conservation of built heritage, but the main one is concerning how original uses can be changed while still conserving the importance of the area and its buildings [11]. Conservation aims to be part of a broad set of procedures for promoting the existing physical built environment and affects all citizens in a community. Architectural conservation is something that embraces different forms and subjects such as urban design, housing, environmental issues, and renewal [12]. Built heritage can also create conflict between issues unrelated to built heritage when applying conservation methods. Undeniably, conserving built heritage is a complex process that requires consideration of socio-economic and environmental dimensions [13].

Culture is a necessity and can play an essential role in conserving Baghdad's city centre's urban heritage. People's interest in dwelling in historic city centres is not only due to socio-economic factors but also different activities and the various cultural and urban strata [14]. The challenge in Old Rusafa is whether urban heritage conservation plays a significant role in promoting social life and motivating new generations to move back to the city centre. The key question is how can the preservation of the physical built heritage sustain conservation strategies in light of multiple challenges: poor management, deteriorating built heritage, and migration? Heritage conservation in Baghdad is impacted by different elements such as architectural, urban planning, socio-cultural, economic, and property rights aspects. These aspects have been identified as essential elements in "heritage disputes surrounding proposals to develop built-heritage properties, and as such, they are instrumental in grounding a conceptual frame" [13]. Conserving Baghdad's built heritage can play an important role in promoting the city's aesthetic value, identity, cultural value, significance of place, as well as economic and commercial value [15]. It also develops the city's historical physical environment and ensures its continuity as an attractive place to live. Urban conservation has become a policy of redevelopment in historic places and social-economic advancement in various cities in the world, such as Johannesburg in South Africa, Penang in Malaysia, Singapore, and Hong Kong [16]. Therefore, urban conservation in Baghdad's old centre requires comprehensive spatial analyses and investigations devoted to the evaluation of urban historic areas [16–18].

During the 19th and 20th centuries, the traditional compact urban fabric and cultural heritage in Baghdad suffered from poverty, migration, overpopulation, unemployment, and segregation [19]. However, historical buildings going back to the early 13th century, such as the Al Mustansiryia School, the Abbasid Palace, and the Al Khulafa Mosque, have all resisted these challenges and new forms of transformation. In the 20th century, the Municipality of Baghdad, due to a lack of expertise, financial constraints, wars, civil wars, and unstable political situations, did not consider the importance of revitalising the built heritage that represents the city's social, economic, and environmental values embedded in its heritage. Consequently, this has led to irreparable damage to the built heritage of the city. Many heritage buildings have been neglected and pulled down in Baghdad without clear conservation strategies to conserve the city's built and cultural heritage.

1.3. Sustainable Built Heritage and Conservation

The beginning of international initiatives to combine both sustainability and conservation was in 1972 by the United Nations (UN) and UNESCO. These initiatives did not emphasise the words sustainability and conservation at that time, as the emphasis was primarily on the environment and protection [20]. The concept of sustainable built heritage has been examined in ways to investigate the advantages of the integration of sustainability principles and built heritage conservation processes and policies [21–24]. The built heritage can be considered a main aspect of sustainable built environment management by participating in the domains of socio-economic and environmental sustainability [25].

The urban environment in Baghdad has been faced with unprecedented change in the last century. Many elements are leading to change in the urban environment, such as demographic changes, globalisation, uncontrolled development, and economics, which directly affect the conservation of historic urban environments [26]. Therefore, the conservation of Baghdad's built heritage can offer sustainable solutions to many problems that afflict traditional areas, such as social, economic, and environmental problems [27]. Sustainable urban heritage in Baghdad can be accomplished through creative ideas and community engagements in the city's future urban planning [24].

1.4. Built Heritage Conservation and Citizen Participation

In recent years, governments, practitioners, experts, academics, and international organisations have considered heritage as urban areas rather than single monuments. They moved towards utilising citizen participation to preserve, manage, and control urban conservation plans. The traditional urban fabric in historic cities in developing countries suffers from similar cases. Firstly, they are currently facing fast population growth, poor infrastructure, and demolitions of urban heritage. Secondly, and most importantly, they ignore the significance of citizen participation and opinion as an essential element to resolving problems that might appear in the decision-making and policy-making processes. The bottom-up approach has become an essential factor in cultural heritage issues, creating new opportunities for citizens' participation in the decision-making process of urban conservation. Citizen participation can improve the practice of urban conservation and assist policy-makers in identifying opportunities and challenges [28].

Nowadays, enhancing the concept of identity and traditional social values is a tendency in Old Rusafa. Baghdad's old centre is represented by the important role of culture, religion, and the most significant element of social structure. The physical and social situations are depressed in this traditional area; however, it is still the source of cultural inspiration for citizens, and people still prefer to go back and live in the traditional neighbourhoods where they grew up. This area has community representatives who could participate in and promote conservation processes more than the official administrative body in such areas [27]. Politics and top-down decision-making issues in Baghdad make it more difficult to formulate a plan for urban modernity that is consistent, specific, and unanimously accepted [29].

The Iraqi Government and the Municipality of Baghdad, between 1956 and 2011, appointed various international architects and planners, such as P.W. Macfarlane in 1956, Doxiadis in 1958, Polservice in 1973, and JCCF in 1987, to prepare Baghdad's future development master plan. However, these initiatives did not consider citizens' participation as a main element in the city's sustainable future development [5,27]. These plans have concentrated instead on demolishing big parts of the built heritage in the city and on simulating Western countries' methods of urban growth. Consequently, Baghdad's old centre-built heritage, traditional urban fabric, and heritage buildings have deteriorated and been left without any form of preservation strategy, with many heritage properties being demolished to make way for contemporary buildings that have little cultural relation to Baghdad's culture and identity. Academic research has urged decision-makers to promote public consultations in the field of heritage, yet this strategy has not been extensively applied in Baghdad. Active community engagement in preserving heritage can enhance sustainable urban heritage development [30]. Thus, the Municipality of Baghdad should consider Baghdad residents' participation or a bottom-up system in the city's future development plans if they want to conserve the city's urban heritage sustainably.

1.5. Conservation Heritage in Arab Cities

To understand existing challenges and possible tools to be used in heritage protection, a comparative perspective can be useful. Comparison in heritage studies can help explore existing problems in a larger context while overcoming the problem of policy insularity [15,31,32]. Thus, lessons gleaned from nearby Arab cities are of importance.

Arab cities, especially in the historical city centres, share some similarities, such as compact urban fabric, narrow passageways, and courtyard houses. Despite their differences in geographical topography, climate, and location, they share the same socio-cultural characteristics and local construction materials and systems (Figure 2) [33]. The Islamic urban heritage that formed Arab cities' heritage centres faced many challenges of modern development and destruction of the heritage-built environment. Most of these historical centres are rich with many heritage buildings and traditional architectural values that have been destroyed and replaced by modern high-rise buildings [34,35]. The owners of heritage buildings lost interest in preserving their properties and moved out to new modern areas, as can be seen in the cases of Baghdad, Cairo, Tunis, downtown Beirut, and Damascus [36]. Many properties in Baghdad's city centre were abandoned, and some were expropriated and confiscated by the state following the hollowing out of the city centres, the out-migration of a huge Jewish community in the 1950s, and Christians and Muslims throughout the 20th century. Thus, improving the quality of urban areas in recent years has become a response to the challenges displayed by citizens' mobility, which require the physical renewal of declining inner urban spaces flexibly.

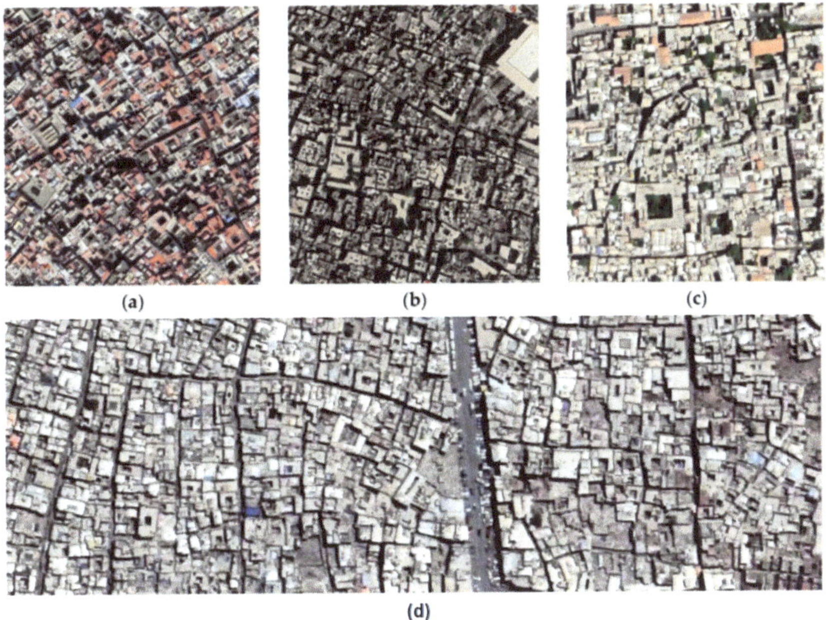

Figure 2. Arab cities' historic urban fabric. Four cities having the same inner-oriented openings and narrow irregular shaded pedestrian networks: (**a**) Rabat historic urban fabric; (**b**) Cairo historic urban fabric; (**c**) Damascus historic urban fabric; (**d**) Baghdad historic urban fabric. Source: author 2024, according to [33].

In the Arab world, many countries have witnessed a spectacular rapid urban evolution after gaining independence in the 1950s and 1960s. Arab cities' municipalities such as Cairo, Tunis, downtown Beirut, Damascus, and Sana'a have shown that they cannot afford to preserve and rehabilitate the majority of their urban heritage and monuments [36]. Thus, international initiatives proposed to preserve some heritage in Arab cities such as Baghdad, Cairo, Tunis, downtown Beirut, Damascus, Fez, and Aleppo by their city governments (Figure 3). Studies by international organisations, such as the United Nations (UN) and UNESCO, the World Heritage Committee (WHCom), and ICOMOS, indicate that Arab cities have lost some of their built heritage and require an urgent sustainable conservation plan to protect the reset [36–38].

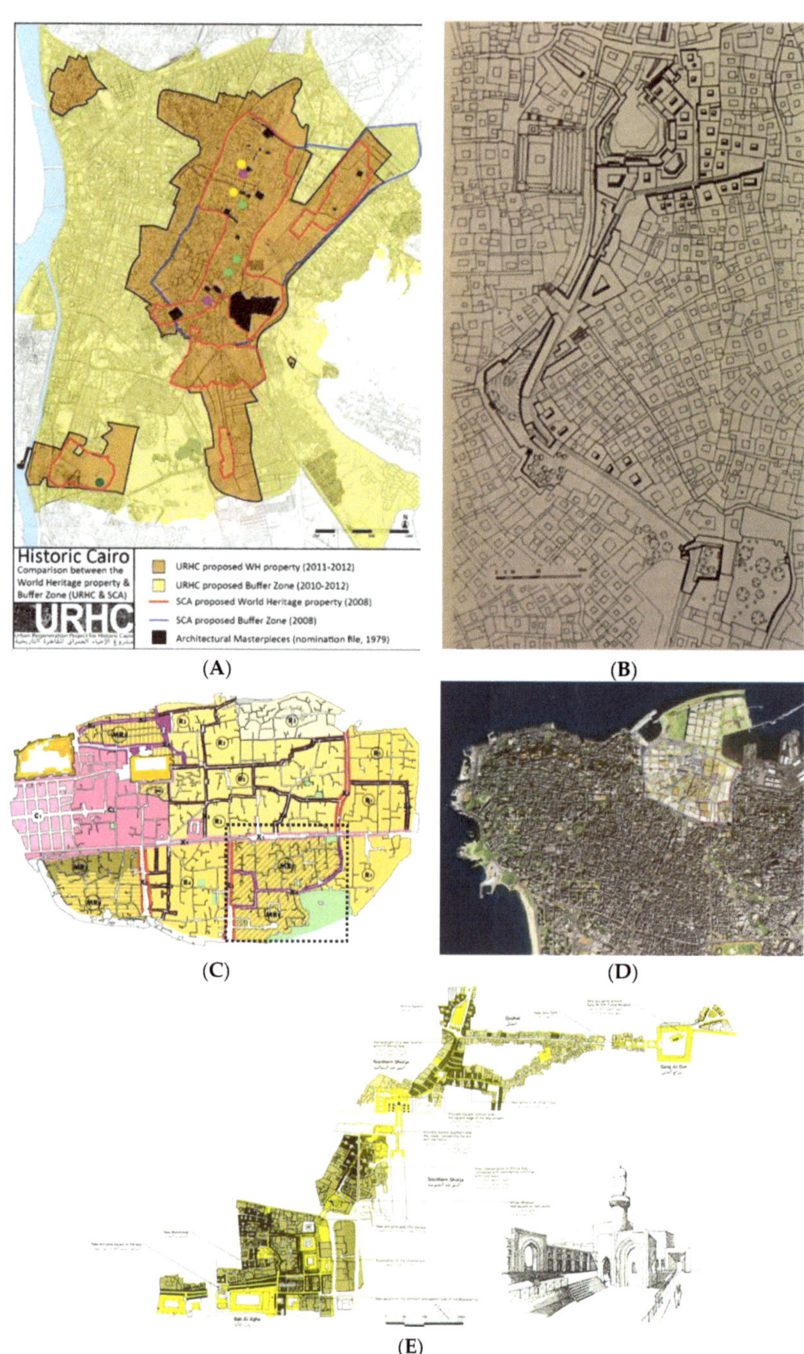

Figure 3. Arab cities rehabilitation plan. Source: author, according to [37–41]. (**A**) Historic Cairo rehabilitation plan. Source: [37]. (**B**) Fez rehabilitation plan. Source: [38]. (**C**) Old City of Damascus plan. Source: [40]. (**D**) Solidere master plan: Beirut. Source: [41]. (**E**) Conservation of main souks of Old Rusafa, Baghdad, JCP Plan 1984. Source: [39].

Cultural heritage and intangible heritage in historic Arab cities have suffered from many problems in recent years, such as fast population growth, poor infrastructure demolition and neglected urban heritage. Arab cities' city councils ignore the significance of citizen participation and opinion as an essential element in resolving problems that might appear in the decision-making, policy-making process and urban heritage conservation. Many built heritage conservation plans prepared to conserve these traditional areas were neither successful nor sustainable due to the narrow scope of urban heritage projects, lack of funds and unsuitable conservation methods adopted. Urban conservation methods that have been considered to face these challenges did not participate in conserving Arab cities' heritage context professionally and sustainably [33].

The process of physical growth in most historic Muslim cities over the past eight decades was determined by the colonial powers in setting out their new modern area. The possible range of urban interventions was defined by two extremes "One consisted in superimposing the new city on the old historic fabric by cutting out large new roads and sites for major public buildings-an approach which entailed the progressive demolition of historic urban structures by the expanding new facilities. The other one consisted of setting up completely new colonial cities on virgin land, without seeking any interface with pre-existing urban structures" [38] (p. 177).

The context and the abovementioned challenges in preserving the built heritage of Arab cities, also explain the multiple hurdles faced by decision-makers and preservationists in Baghdad. A first step towards securing the future of heritage in Baghdad is mapping the existing inventory of heritage structures in the urban historic centre. In addition, it is essential to understand existing policies and heritage plans, their strategies, the cultural elements they protect, and the instruments they utilise to achieve sustainable and long-term heritage protection.

2. Assessing Baghdad's Urban Heritage

The capital, Baghdad, is the largest city in Iraq and contains four historic areas: Old Rusafa, Al-Karkh, Al-Adhamiya, and Al-Kadhimiya (Figure 4). The area of Old Rusafa represents the main historic centre and is an integral part of the central business district located on the eastern bank of the Tigris River. The importance of the historical centre is not only local but also regional and national dimensions. The old centre comprises the largest concentrations of historical workshops and souqs, as well as some of the important mosques and government buildings in the country. The main significant features in the historic part of Baghdad are narrow alleys, natural shading, human-scale patterns, high density/low-rise living, the hierarchy between public and private space, mixed-use, a walkable, and a zero-carbon environment that provides an extraordinary base towards achieving sustainable urban heritage. Old Rusafa contains several significant historic buildings: 132 monuments are listed, twenty-one of which belong to the Abbasid Empire (762–1258) and the rest to the Ottoman Period (1638–1917). Therefore, it is considered an important heritage that demands emergency protection for its historical value [39].

The old core of Baghdad is a unique traditional area, and historic components that survive to this day consist of important heritage buildings, spines, and monuments in clusters that show the traditional urban sense of the city (Figure 5). The heritage value represented by the traditional urban fabric contains many significant mosques, such as the Morjan Mosque, which was built in 1356 AD; monumental buildings, such as Al-Mustansirya School, Abbasid Palace, and churches, including St. Joseph's Cathedral, Jewish heritage schools, such as Masooda Saliman School for boys, and traditional markets such as Al-Sharjah Souk (Figure 6). The traditional Baghdadi house is another type of historic building in the old centre. These houses are now squeezed into the inner part of Old Rusafa, between the service and industrial parts of Sheik Omer Street and the predominantly commercial and business part of Al-Rashid Street.

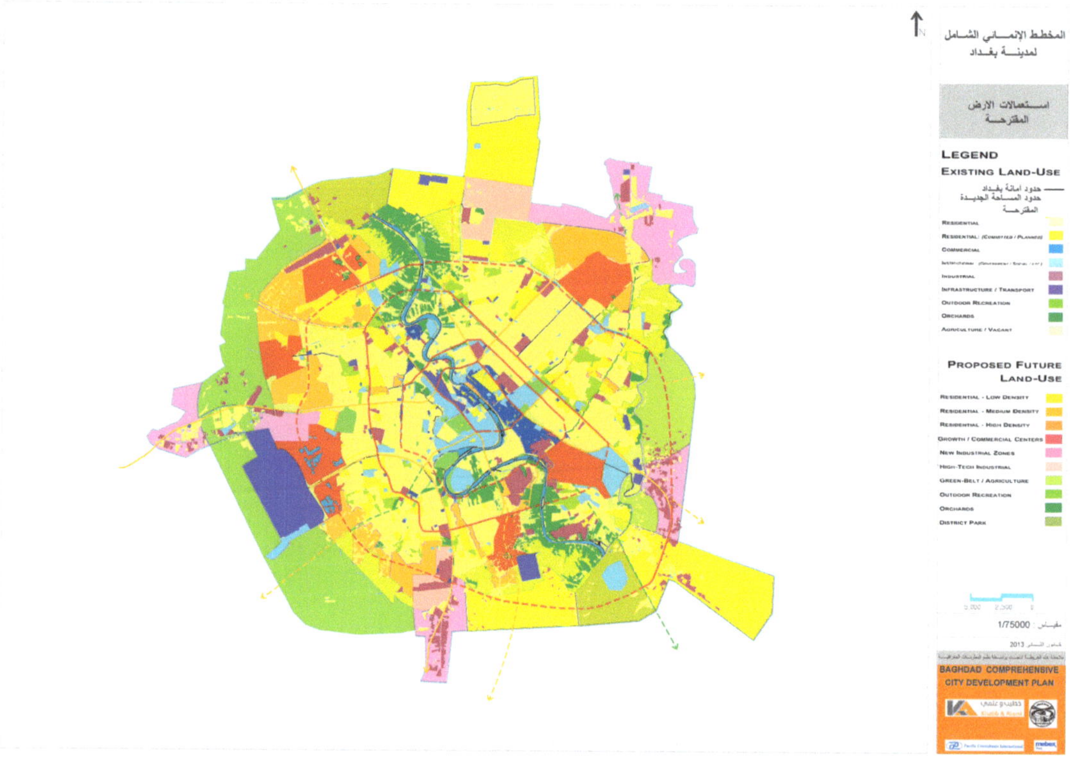

Figure 4. Baghdad's comprehensive development plan for 2030. Source: author, according to [10].

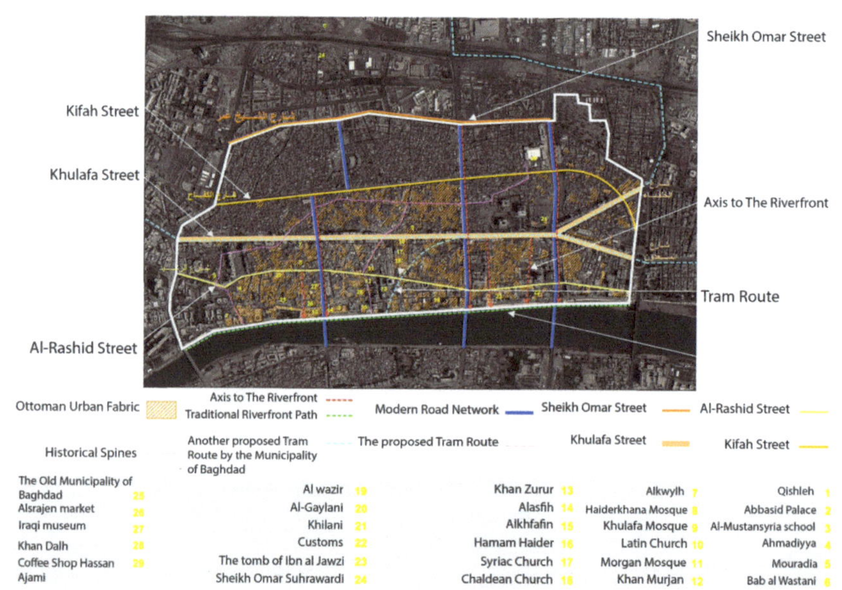

Figure 5. Built heritage in Old Rusafa. Source: author, according to [42].

Figure 6. Built heritage in Old Rusafa. Source: author's original images. (**A**) Morjan Mosque (1356 AD). (**B**) Al-Mustansirya School (1227 CE). (**C**) St. Joseph's Cathedral (1643–1848). (**D**) Traditional souks. (**E**) Traditional Baghdadi house. (**F**) Traditional urban fabric-form.

During the last few decades, Old Rusafa has suffered both from its monuments and historic neighborhoods; however, enough fabric remains to evoke its past grandeur. The JCP Master Plan in 1984 was a significant plan prepared to preserve Old Rusafa urban heritage (Figure 7). The comprehensive surveys by JCP in 1984 showed that the historic area contains 3900 houses, which mostly belong to the late 19th and early 20th centuries, sixty-three mosques, five tombs, six madrassas, eleven khans, six hammams, four churches, nine souqs, and three gates. Of the listed 132 monuments in the old area of Rusafa, only 21 monuments belong to the Abbasid Empire (762–1258), while the majority are from the Ottoman Period (1638–1917) [39]. The *State of Iraq Cities Report* (SICR) 2006–2007 reported that "the structures of historic areas have been modified, with many buildings evolving towards commercial and government use. The city has many seriously deteriorated structures without infrastructure, leading to poor internal sanitation, drainage problems and effluences" [27]. This has led to the migration of many families from the old centre into new, modern areas in Baghdad. More than 50% of all buildings in the case study were in poor or very poor structural condition. The main reason for the deterioration of the physical and environmental heritage conditions of Old Rusafa is the insufficiency of a comprehensive master plan that can adapt sustainable urban heritage design strategies. Moreover, high population densities, clearances for new roads, neglect of the historic urban fabric, and the ongoing conflict and political instability situation in Iraq, particularly in Baghdad, have all led to the built heritage deterioration (Figure 8) [39].

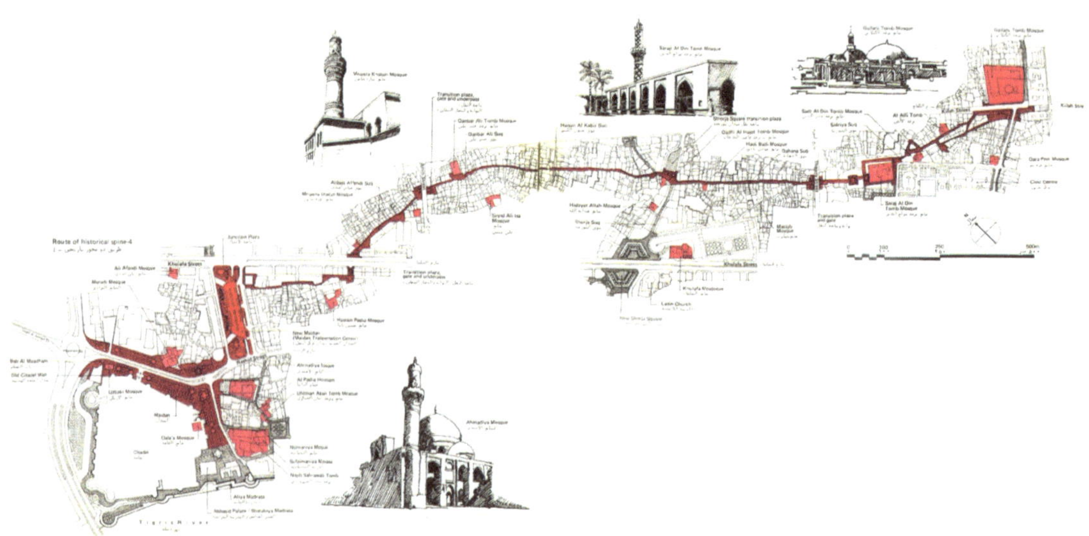

Figure 7. JCP plan in 1984 to preserve one of the historical hubs in Old Rusafa. Source: [39].

Figure 8. Deterioration of the physical built heritage in Old Rusafa. Source: author's original images.

Modernisation in Old Rusafa

The beginning of modernisation in Old Rusafa started in 1869, when the Ottomans demolished the old city walls that were built during the Seljuk rule (1052–1152 AD) and constructed the first residential extensions. The layout of the old city did not change much between the Seljuk period and the end of the 19th century. Many problems have led to the deterioration of the structures of the historic centre that were constructed of brick and timber and had to be rebuilt periodically due to frequent flood damage and fire. The opening of four major roads between 1914 and 1956, which linked the northern and southern ends, has led to the segregation of the continuous urban fabric into isolated fragments. Another problem was the rehabilitation of the disrupted urban form on both sides of these new roads that were given over to wholesale redevelopment. Huge areas of old urban fabric were demolished, and the rest of the traditional urban fabric was ignored. The method of transforming the city centre from the traditional to the modern has been affected not only physically but also socially by the departure of the original people of Old Rusafa into new modern areas. A new community from rural areas searching for employment and a better life filled the social vacuum in the old city by renting traditional houses, usually one family per room. Furthermore, property owners were not interested anymore in preserving their properties. These circumstances have led to the acceleration of physical deterioration.

Consequently, in most of the traditional areas in Baghdad, especially the area of Old Rusafa, the younger generation is abandoning these areas due to the lack of standard infrastructure, the deteriorating built environment, rundown houses, air pollution, and a lack of modern facilities. Thus, the traditional centre needs significant conservation processes under new sustainable methods to preserve city culture and urban heritage (Figure 9).

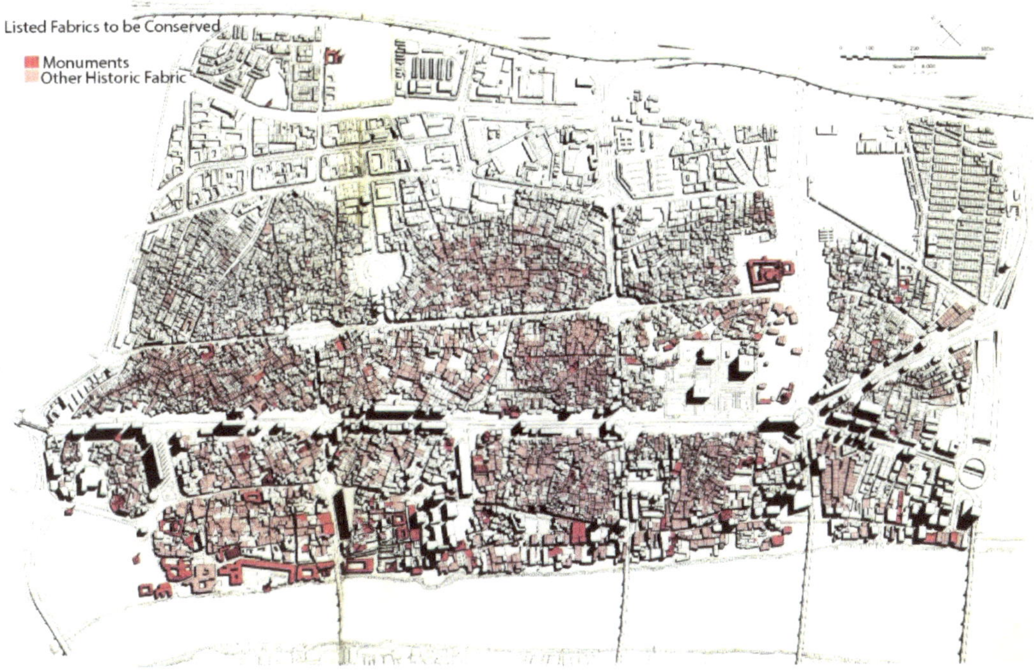

Figure 9. Listed fabrics to be conserved in Old Rusafa. Source: [39].

3. Assessing the Old Rusafa Historical Centre: The Area between Rashid Street and the Riverfront

Old Rusafa is the historic centre of Baghdad, and due to that, its fabric has been under pressure from modern development and has suffered tremendous losses in its traditional form. However, there is still an opportunity to preserve the rest of the unique fabric by promoting these areas with new facilities and fixing broken structures (Figure 10). Thus, this research investigates the historic centre of Old Rusafa as a case study area, which can produce a comprehensive understanding of the need for sustainable urban heritage development. Fieldwork was the method used to gather information for this research. The data was collected from diverse sources associated with the various municipality departments of Baghdad, the Presidency of the Mayoralty of Baghdad, the Baghdad Heritage

Department, the Urban Planning Department of Bagdad, the University of Technology, and the University of Baghdad [4,6,9,43].

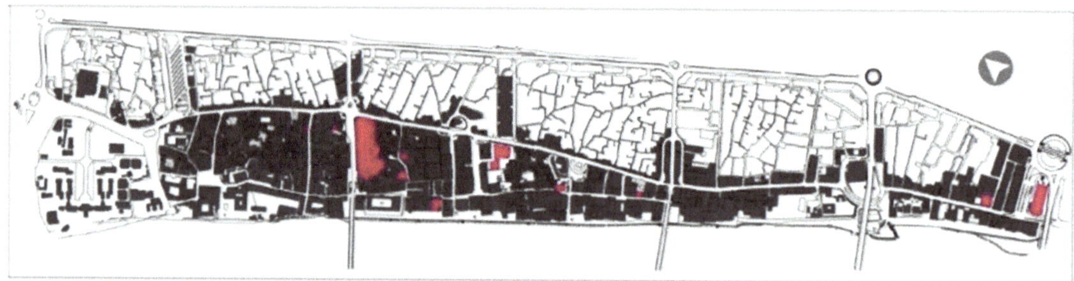

Figure 10. Heritage urban fabric of the area between Al-Rashid Street and the Tigris riverfront in Old Rusafa between 1983 and 2009. Source: Author 2024, according to [44].

The area between Al-Rashid Street and the Riverfront is a very important part of Old Rusafa that might pose both conservation and development problems. The Municipality of Baghdad prepared several proposals to rehabilitate the area, which include detailed assessments of the existing architectural heritage and general design guidelines. One of the significant comprehensive urban conservation master plans for Old Rusafa was submitted by JCP (Japanese planners, architects, and consulting engineers) in 1984. This research examines the area between Al-Rashid Street and the Tigris Riverfront in Old Rusafa. The area consists of a longitudinal strip extending approximately four kilometres between Bab Al-Moatham and Bab Al-Sharqi; therefore, the research focusses on assessing Zone A's built heritage as an example of other zones in the area [2] (Figure 11).

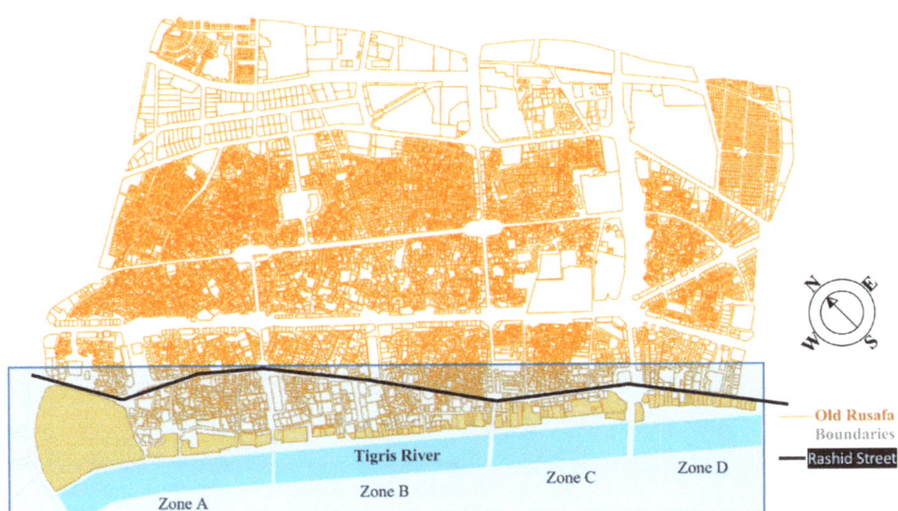

Figure 11. The area between Rashid Street and the riverfront in Old Rusafa. Source: author's original image.

3.1. Historical Background of Al-Rashid Street and Squares

In the late 19th century, the Ottomans began to cut the first axis in the traditional urban areas and attempted to import Westernisation into the urban evolution pattern [45]. The opening of Al-Rashid Street in 1915, completed by the British in 1917, was the first change in the traditional urban fabric [46]. The traditional-style buildings on Al-Rashid Street had a consistent type, which gave the street a continuous character (Figure 9). However, its unique character is now threatened by numerous high-rise office blocks, traffic jams, parking on the street, and the neglect of its fabric [45]. The street became the most significant feature and the centre of business in Baghdad for decades.

In the urban planning scene, Al-Rashid Street has not only been considered the first modern street in Baghdad but also in Iraq. The traditional street is about 3120 m long and about 12 m wide. It was started by Khalil Pasha in 1915 and completed by the British around 1917. The street linked the two old gates of Old Rusafa and was characterised by its colonnaded paths, traditional two-floor buildings, human scale, architectural unity, and convenient shading paths for pedestrians. The traditional buildings and the variety of architectural styles on both sides of Al-Rashid Street show the story of the development of the old city over the last one hundred years. Its building can be characterised by the workmanship and traditional materials, which have been quite remarkable assets to good design and detailing (Figure 12). The arcade of these traditional buildings along both sides provides a colonnaded walkway approximately 3.5 m wide and 5 m high. Physically and morphologically, this street is divided into four main areas due to the construction of bridges from 1939 to the present that resulted in physically cutting Al-Rashid Street into these parts. Nowadays, there are several highly modern buildings that penetrate the traditional street of Old Rusafa, ignoring the unique character of the street and confusing the urban scene of the historic centre. This has been happening since the 1950s because of a lack of efficient urban design policies and urban development controls. Al-Rashid Street has represented the commercial and cultural centre of Baghdad for many decades, and today it is a part of the CBD area of the capital city [39].

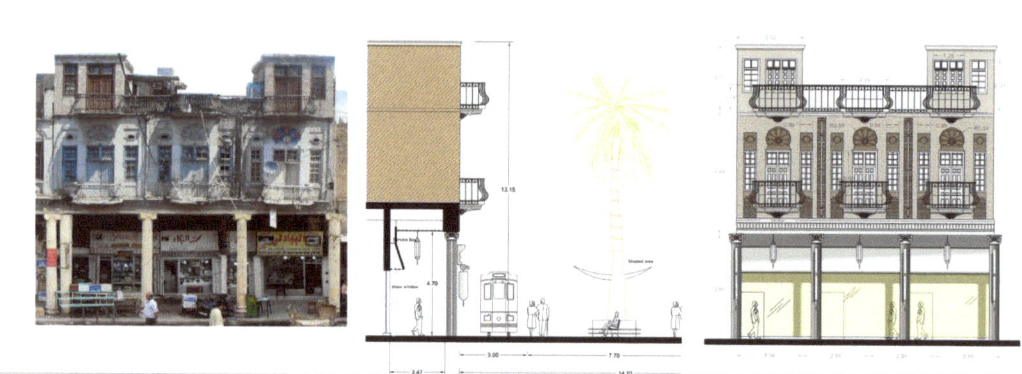

Figure 12. Sustainable future development initiative in Al-Rashid Street. Source: [47].

Many significant historical and traditional buildings are located in Al-Rashid Street Zone A, such as Haydar Khana Mosque, which was built by Dawood Pasha in 1827 AD (Figure 13). Heritage buildings in Zone A require short- and long-term sustainable conservation strategies to preserve their heritage as they suffer from deteriorated structures and poor maintenance. In Zone A, several important heritage buildings have unique architectural styles that represent the street and city's cultural heritage. The buildings' heights along Al-Rashid Street, on both sides of Zone A, are between two to three floors with a shaded arcade [2] (Figure 14).

Figure 13. Al-Rashid streets A historical buildings and Haydar Khana Mosque. Source: author's original images.

Figure 14. Al-Rashid Street architectural styles in Zone A. Source: author's original images.

3.2. Historical Background of Tigris Riverfront

The Tigris River is one of the most important natural aspects that have played a significant role in promoting the prosperity of Baghdad from the past centuries until now. It has been considered the main source of irrigation and development in the surrounding agricultural areas of the capital city of Iraq; furthermore, it has become a space of entertainment and an element of connection with other cities. Three boat bridges crossed the Tigris River and connected Old Rusafa and Al-Karkh during the late Abbasid period (1055–1258), when Old Rusafa expanded into its familiar rectangular form [39].

The morphology of Old Rusafa was influenced by the Tigris and all-important functions that were located on its bank or very close to it, such as the Citadel, traditional souqs, mosques, and monumental buildings. The riverfront is divided into separate areas with various architectural features [48]. However, the riverfront nowadays is almost neglected, and its character has been disturbed by numerous tall buildings that have been built near some magnificent buildings, such as the Al-Mustansiriya School, which was built in 1227.

The Old Rusafa Tigris Riverfront historical buildings may be an attractive area that plays an essential role in promoting city culture and socio-economic and environmental aspects. The Riverfront heritage requires implementing sustainable urban heritage design strategies to conserve its built heritage.

Observation Survey for Tigris Riverfront Zone A

In Tigris Riverfront Zone A, a combination of the observation, walking and serial vision methods has been adopted as a qualitative approach for gathering information to evaluate the existing physical urban heritage context and form (Figure 15). In Figure 15A, the Ministry of Defence Building is one of the significant buildings in the area that also represents the old citadel area. During the civil war in 2007, the building was surrounded by a concrete wall to prevent any outside attack, this has led to the separation of the riverfront and the destruction of the continuity and walking activity along the riverbank. The next heritage building to the Ministry of Defence buildings is the Abbasid Palace built between 1180 and 1225 AD (Figure 15B). The traditional building is neglected and suffers from poor conservation of its architectural heritage features and identity, even the area and open spaces between the Abbasid palace and the Tigris Riverfront show poor design strategies that could be developed to attract visitors and tourists and influence the area's socio-economic and environmental aspects. Another important building on the riverbank that is called the traditional school building, its white elevations and structure have been well preserved (Figure 15C).

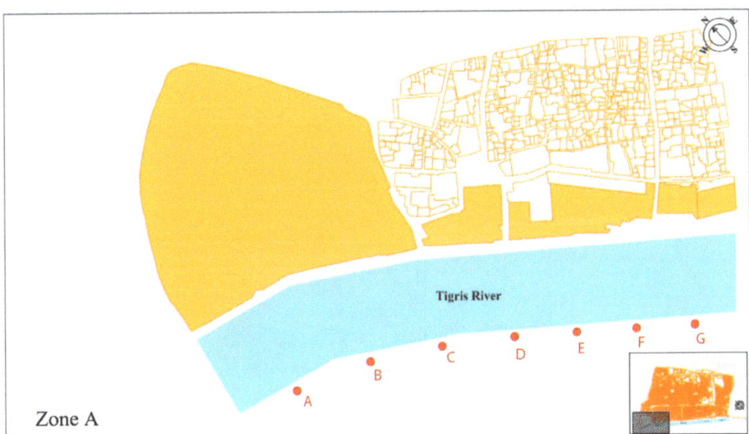

Figure 15. *Cont.*

Figure 15. (**A–G**): Observation survey for Tigris Riverfront Zone A. Source: author's original image.

In recent years, a new building was constructed with yellow brick and arched windows (Figure 15D) next to the heritage-listed building called the House of the Governor, which is connected to a mosque from its left side. This historic building is another building that requires intensive conservation work for its architecture and built heritage (Figure 15E). The famous Al-Qishla Clock Tower is surrounded by the Al-Sarai heritage building, both buildings were built in 1869 (Figure 15F). Cultural Baghdadi Centre and Al-Wazir Mosque were two main traditional buildings at the end of Zone A. Al-Wazir Mosque was built by Hassan Pasha in 1600 and rebuilt again by the Awqaf in 1957 after being ruined by the severe floods of 1831 [2] (Figure 15G).

The outcomes of riverfront observation reveal that no sustainable conservation strategies have been adopted to preserve the area's open spaces, riverbank, and most importantly, the built heritage. Many heritage buildings in Zone A suffered from poor structural conditions and require urgent architectural feature maintenance. Therefore, Baghdad and the Old Rusafa area especially require a future development plan that adopts sustainable urban heritage conservation strategies to protect the city-built heritage sustainably.

3.3. Historic and Architectural Value in Zone A (Architectural Types)

Historic and architectural value is one of the significant elements in the examination of the case study area to achieve sustainable indicators. Therefore, Zone A buildings have been examined in the field investigation according to their historical and architectural importance, dividing them into seven categories (historic buildings, early traditional, art nouveau, art décor, modern buildings, empty spaces, and unknown) (Figures 16 and 17). The vast majority (45%) of five hundred and fifty buildings have no architectural style according to the outcomes of the field survey of the historical and architectural value in Zone A, whereas historical buildings belonging to the Abbasid Empire (762–1258) were recorded at 10% in this survey. The traditional buildings related to the Ottoman Period (1638–1917) accounted for 13% of all buildings in the case study area Zone A (Figures 16 and 17). The results also asserted that modern buildings in Zone A were recorded at 11%. Art Nouveau (Art Nouveau is the type of building constructed with brick and steel I sections going back to the late 1920s and early 1930s and is characterised by its organic lines that were influenced by European style that began to reach Iraq at the time) was another important element that was assessed by the field survey of the old centre of Baghdad, which accounted for 5% of all buildings in Zone A. Art Déco is another category in Old Rusafa; this category's buildings are constructed with brick and steel I sections, with a geometric decoration of the brick material and use of iron material in the balconies. This type is characterised by straight lines and angles in the architectural elements of its elevations, going back to the 1930s and early 1940s), and accounted for 4% of this examination. The field survey also shows that empty spaces were recorded at 12% in Zone A of Al-Rashid Street [44,47].

Figure 16. Historic and architectural value in the Zone A (architectural types). Source: author, according to [44,47].

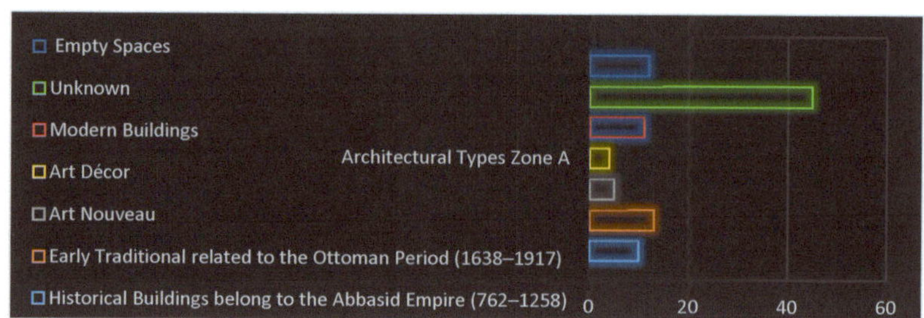

Figure 17. The percentage of historic and architectural building types in Zone A. Source: author, according to [44,47].

3.4. Historic and Traditional Buildings in Zone A

This field examination of Old Rusafa Zone A revealed that there were thirty-five buildings considered traditional buildings within Zone A, denoting a particular use (Khan Al-Mdallal and Al-Rasheedia School). Moreover, this part of Al-Rashid Street has included fourteen historical buildings, denoting a particular use (the Saray building and its Qushla clock tower, which was built in 1869; the Al-Srajeen Souq was built in 1802; and the court zone) (Figure 18). Zone A contains several significant souqs, such as Mutanabi Street, which includes the book market, and Saray Souq, which sells stationery. This zone of Old Rusafa has embraced many traditional mosques, such as the Ahmadiya mosque, which was built by Ahmad Pasha in 1769; the Numaniya mosque, which was built in 1772 by Fatima Biktash al-Sayid Wali to memorialise her husband, Numan Agha Ibrahim; and the Saray Mosque, which was built in 1704 by Hasan Pasha. The Zone A district, as it was asserted by this survey, contained significant traditional buildings, such as the Baghdadi Museum, which was built in 1910. Moreover, this part of the case study area includes traditional Baghdadi houses like the Kihami house, which was built in 1920, and the Salima Daud house, which was built in 1900 [44,47].

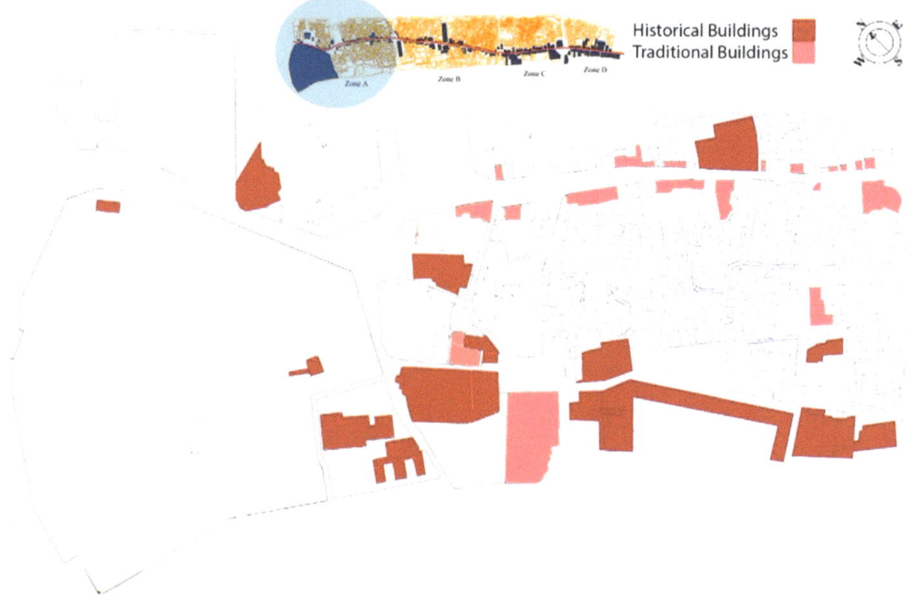

Figure 18. Historic and traditional buildings in Zone A. Source: author, according to [44,47].

4. Sustainable Urban Heritage in the Historic Centre of Baghdad

Baghdad's urban heritage areas are characterised by human scale, natural shading, privacy, mixed-use, walkable, natural environment, and organic narrow alleys with a homogeneous arrangement of housing plots (Figure 19). Old Rusafa is well-defined by its unique traditional urban fabric is and surrounded by a modern urban pattern and roads, which replaced their walls. The features of the old urban fabric represent the main principles of environmental sustainability and show how individuals in the past built a sustainable environment to face the tough climate. Through their design, they achieved equity, a clean environment, an efficient use of resources, safety, low use of energy, and a low rate of pollution. Sustainable urban heritage in Old Rusfa should be able to provide appropriate solutions to regenerate the traditional fabric in terms of urban form, land use, local environment, and transportation, as well as create a new vision to deal with socio-economic and environmental processes. Based on existing scholarly publications, it is possible to assert that sustainability can become an essential strategy for heritage protection. Preserving Baghdad's urban heritage can aid in sustainable solutions to many problems that afflict the historic core, such as socio-economic and environmental issues. Achieving sustainable urban heritage in the city can participate in developing people's quality of life, promoting city diversity, and enhancing the old centre built and cultural heritage environments at local and regional levels [20]. For example, the preservation of old and narrow alleys can contribute to the city's sustainability by providing shade to pedestrians and human-scale development.

Figure 19. A narrow alley with natural shading in the historic part of Baghdad. Source: author's original image.

The main challenge for architects, urban designers, and policy-makers in conserving Baghdad's built heritage is to develop a sustainable urban heritage framework and design strategies for urban heritage conservation. To achieve this, the Municipality of Baghdad needs to integrate socio-economic and environmental aspects into sustainable urban heritage conservation to improve Baghdadi people's quality of life and develop a better place to live and work [49].

5. Discussion

Urban heritage conservation is "increasingly perceived as a generator of income which extends to its function as a tourist attraction and item of leisure consumption, and this constitutes the dilemma of dissonant heritage. Very often, this leads to conservation and revitalisation projects having conflict with the cultural role of heritage and loss of social continuity" [49]. Many challenges emerge in the conservation of built heritage, but the main one is that the original uses may change while preserving the importance of the area and its buildings, which are related to the area's new economic development projects that led to deculturing and destruction of many cities' urban heritage. In Old Rusafa, for example, several listed buildings have been transformed from their original land use into new uses. Many listed Khans have transformed into restaurants and cafes, and several traditional Baghdadi houses have been transformed into art centres and museums. As a result, the area's heritage identity and land use characteristics have been changed, which has impacted the area's social aspect.

Maintenance and repair of community properties are one of the traditional systems that have been used by Islamic society, and this was established within a kind of endowment called waqf. This system was a result of the relation of Islamic philosophy to equity. The waqf system depends on voluntary contributions to administer properties such as public and social services, mosques, caravanserais, and schools. This system has continued through to modern times in many Islamic countries, such as Iraq, Morocco, and Egypt, guaranteeing the repair of traditional buildings and preventing the division of larger properties between several inheritors, and has laid the groundwork for common social responsibility [50]. The Waqf system in Baghdad was established in the time of Harun Al-Rashid as a system to maintain and preserve important city places and buildings. Nowadays, waqf plays a limited role in maintaining or preserving its own heritage and modern buildings in Old Rusafa. The Municipality of Baghdad has taken on this role to prepare and manage conservation plans for Baghdad's built heritage.

One of the main challenges in the case study of this research today is how to find a platform that could solve the battlefield between various architectural styles and ideologies. The lack of a comprehensive conservation master plan to preserve Old Rusafa built heritage has led to a conflict between, on the one hand, the modern development that shows no consideration for the traditional urban context and, on the other, the large stock of abandoned, decaying, or misused traditional buildings in the case study area. The main causes of the present lack of consistency in traditional urban fabric are as follows:

- Urban growth that has shown little opportunity for continuity and organic expansion.
- Insufficiency of a comprehensive master plan that can adapt urban design methods related to the special features of the historic core.
- The new architectural evolution projects that have been built in Old Rusafa have ignored and isolated the existing traditional urban fabric.
- There is a lack of clear conservation methods that can implement the renovation, preservation, rehabilitation, and promotion of traditional buildings.
- The lack of clear policy and development schemes that would improve and give a positive motive to advance methods that would be applicable in other locations.
- The ongoing conflict and political instability in Iraq, particularly in Baghdad.

The Municipality of Baghdad lacks a clear vision and regulations regarding urban conservation, and this usually creates many obstacles when they want to prepare a plan for conserving traditional areas. The heritage urban fabric in Old Rusafa has witnessed irreparable damage due to the weak definition of functions and an ambiguous formulation of what to preserve, political issues, top-down systems, and wars. These are some of the reasons why most urban conservation plans prepared by different groups for the city centre have not been implemented successfully, such as the initiative prepared by the Union of Architectural Heritage in 2010 to rescue Baghdad's architectural heritage [27].

Therefore, to create an effective sustainable urban conservation plan to protect Old Rusafa's built heritage, the Municipality of Baghdad, architects, and urban designers should map heritage at risk, unravel its unique qualities that promote sustainability, run a better inventory of heritage assets, and think of sustainable and durable policies (including incentives) that encourage rehabilitation. The Municipality of Baghdad should work closely and develop a collaboration platform with the World Heritage Committee (WHCom), UNESCO, the United Nations (UN), and ICOMOS to protect Baghdad's city heritage. Finally, it would also need to learn lessons from the Arab world countries that have faced many challenges in preserving their city cores, such as Cairo, Damascus, Beirut, Aleppo, and Fez.

The Municipality of Baghdad, Iraqi architects, professionals, and urban planners should develop their attitudes towards preservation strategies and adopt some lessons in sustainable urban conservation from near Arab cities. The Old Rusafa sustainable urban heritage conservation process requires a political environment of development where Old Rusafa is rehabilitated for its heritage value. Furthermore, to achieve sustainable urban heritage preservation in Baghdad's old centre, society and civic organisations should participate in preserving their heritage properties that are not under the Municipality of Baghdad's listed heritage buildings. The Old Rusafa residents must be educated to understand the importance of rehabilitating the area and develop a sense of self-identity to prevent the area from modernisation and gain sustainable urban heritage in Baghdad's city centre.

6. Conclusions

This article investigates how we might achieve sustainable urban heritage conservation in Baghdad's historical centre, Old Rusafa. It has asserted that "conservation, on an urban scale, is concerned with the urban fabric as a whole and not with architecture alone. The successful conservation project will make use of quantitative analyses and will be aided by comparative and economic studies" [14] (p. 13). Thus, the paper argues that the fundamental features of cities demand tangible conservation of their urban fabric to maintain its unique characteristics [16]. Furthermore, the paper illustrated that urban conservation offers a sustainable solution to social and economic problems, promotes the historic environment, creates new opportunities, and brings new life to run-down areas [51].

This research has examined the main challenges in the case study, such as the conflict between modern development that shows no consideration for the traditional urban context and the large stock of abandoned, decaying, or misused traditional buildings that have led to confusion and chaos in Old Rusafa. The paper presented the data gathered by the field survey of the area between Al-Rashid Street and the Tigris Riverfront in Old Rusafa. The field survey was conducted to assess the current situation of the existing physical built heritage of the area between Al-Rashid Street and the Tigris Riverfront in Old Rusafa and define the potentialities and constraints of this area. The field investigation illustrated the architectural value of traditional buildings in the case study area.

Ultimately, the analysis suggests that to tackle the deterioration of existing built fabric, public administrators and decision-makers should adopt a variety of strategies to ensure the long-term sustainability of efforts to regenerate the historic core, including continuous mapping, damage control, finding economic anchors, incentives, and durable policies that can mitigate the longstanding impacts of multiple socio-economic challenges faced by the city.

Funding: This research received no external funding.

Institutional Review Board Statement: Not applicable.

Informed Consent Statement: Not applicable.

Data Availability Statement: No new data were created or analysed in this study. Data sharing is not applicable to this article.

Conflicts of Interest: The author declares no conflicts of interest.

References

1. Al-Saffar, M. Toward an Integrated Sustainable Urban Design Framework in the Historic Center of Baghdad. *Int. J. Environ. Sustain.* **2016**, *13*, 31–52. [CrossRef]
2. Al-Saffar, M. *Toward an Integrated Smart and Sustainable Urbanism Framework in the Historic Centre of Baghdad. (Old Rusafa as a Case Study)*; Manchester Metropolitan University: Manchester, UK, 2018.
3. Global Heritage Fund (GHF). *Iraqi Heritage Experts Call on Government to Preserve Heritage Sites 2012*; Global Heritage Fund (GHF): San Francisco, CA, USA, 2012.
4. Scholz, R.W.; Tietje, O. *Embedded Case Study Methods: Integrating Quantitative and Qualitative Knowledge*; SAGE: London, UK; Thousand Oaks, CA, USA, 2002.
5. Al-Saffar, M. Baghdad: The city of cultural heritage and monumental Islamic architecture. *Disegnarecon* **2020**, *13*, 1–5.

6.   Khalifehei, H. *Social Sustainability and the Future in the Iranian Historic Neighbourhoods' Townscape*; University of Sheffield: Sheffield, UK, 2014.

7.   Leech, N.L.; Onwuegbuzie, A.J. A typology of mixed methods research designs. *Qual. Quant.* **2009**, *43*, 265–275. [CrossRef]

8.   Al-Saffar, M.; Bathsha, A.; Chang, C.-H.; Ge, H.; Huang, X.; Shi, K.; Lu, Y.; Peng, Z.; Sun, W.; Wang, H.; et al. How to Enhance the Future of Urban Environments Through Smart Sustainable Urban Infrastructures? *Disegnarecon* **2019**, *12*, 14.1–14.4.

9.   Wilson, V. Research Methods: Design, Methods, Case Study…oh my! *Evid. Based Libr. Inf. Pract.* **2016**, *11*, 39–40. [CrossRef]

10.   The Municipality of Baghdad. *Baghdad City Comprehensive Development Plan (2030)*; The Municipality of Baghdad: Baghdad, Iraq, 2013.

11.   Al-Saffar, M. Urban Heritage and Conservation in The Historic Centre of Baghdad. *Int. J. Herit. Archit.* **2018**, *2*, 13. [CrossRef]

12.   Glendinning, M. *The Conservation Movement: A History of Architectural Preservation: Antiquity to Modernity*; Routledge: London, UK, 2013.

13.   Mualam, A. Architecture is not everything: A multi-faceted conceptual framework for evaluating heritage protection policies and disputes. *Int. J. Cult. Policy* **2020**, *26*, 291–311. [CrossRef]

14.   Cohen, N. *Urban Conservation*; MIT: Cambridge, MA, USA; London, UK, 1999.

15.   Everard, A.J.; Pickard, R.D. *Can Urban Conservation Be Left to the Market? The Value of Partnership-Led Conservation Regeneration Strategies*; WIT Press: Southampton, UK, 1997; pp. 619–632.

16.   Su, X. Urban conservation in Lijiang, China: Power structure and funding systems. *Cities* **2010**, *27*, 164–171. [CrossRef]

17.   Pendlebury, J.; Strange, I. Urban conservation and the shaping of the English city. *Town Plan. Rev.* **2011**, *82*, 361. [CrossRef]

18.   Koramaz, T.K.; Gulersoy, N.Z. (Eds.) Users' Responses to 2D and 3D Visualization Techniques in Urban Conservation Process. In Proceedings of the 2011 15th International Conference on Information Visualisation, London, UK, 13–15 July 2011.

19.   Selim, G. *Unfinished Places*; Ache, P., Ed.; Routledge: New York, NY, USA, 2017.

20.   Rodwell, D. *Conservation and Sustainability in Historic Cities*; John Wiley & Sons, Inc.: Hoboken, NJ, USA, 2008.

21.   Alexandru, B.; Dorina, C.; Dana Maria, C.; Tudor, B.; Constantin, C.B.; Lucian, C. Heritage Building Preservation in the Process of Sustainable Urban Development: The Case of Brasov Medieval City, Romania. *Sustainability* **2022**, *14*, 6959. [CrossRef]

22.   Étienne, B.; Juste, R.; Georges, A.T. Using sustainability indicators for Urban Heritage Management: A review of 25 case studies. *Int. J. Herit. Sustain. Dev.* **2015**, *4*, 23–34.

23.   Maria, L.; Wouter, V. Sustainability Assessment of Urban Heritage Sites. *Buildings* **2018**, *8*, 107. [CrossRef]

24.   Sanober, N.; Salman, S. The Role of Cultural Heritage in Promoting Urban Sustainability: A Brief Review. *Land* **2022**, *11*, 1508. [CrossRef]

25.   Ragheb, A.; Aly, R.; Ahmed, G. Toward sustainable urban development of historical cities: Case study of Fouh City, Egypt. *Ain Shams Eng. J.* **2021**, *13*, 101520. [CrossRef]

26.   Paul, J. *Historic Urban Environment Conservation Challenges and Priorities for Action*; The Getty Conservation Institute: Los Angeles, CA, USA, 2010.

27.   Al-Akkam, A. Urban Heritage in Baghdad: Toward a Comprehensive Sustainable Framework. *J. Sustain. Dev.* **2013**, *6*, 39–55. [CrossRef]

28.   Tigran, H.; Krister, O. Urban Heritage, Planning and Design and Development. *Sustainability* **2023**, *15*, 12359. [CrossRef]

29.   Pieri, C. Baghdad 1921–1958. Reflections on History as a "Strategy of Vigilance". 2014; Volume 8, pp. 69–93. Available online: https://hal.archives-ouvertes.fr/halshs-00941214/document (accessed on 23 July 2024).

30.   Ferdous, F.; Lawless, J.; Silva, K.D. Sustainable urbanism and urban heritage conservation. In *The Routledge Handbook on Historic Urban Landscapes in the Asia-Pacific*; Routledge: London, UK, 2020; pp. 363–376.

31.   Ryberg-Webster, S.; Kinahan, K.L. Historic preservation in declining city neighbourhoods: Analysing rehabilitation tax credit investments in six US cities. *Urban Stud.* **2017**, *54*, 1673–1691. [CrossRef]

32.   Nir, M.; Nir, B. Evaluating Comparative Research: Mapping and Assessing Current Trends in Built Heritage Studies. *Sustainability* **2019**, *11*, 677. [CrossRef]

33.   Ahmed Mohamed, S. Current Trends in Urban Heritage Conservation: Medieval Historic Arab City Centers. *Sustainability* **2022**, *14*, 607. [CrossRef]

34.   Boussaa, D. Islamic urban heritage: Blight or blessing? In Proceedings of the International Conference on Islamic Heritage Architecture and Art (Islamic Heritage Architecture 2016), Valencia, Spain, 17–19 May 2016; WIT Press: Southampton, UK, 2017; pp. 344–354.

35.   Al-Saffar, M. Assessment of the process of urban transformation in Baghdad city form and function. In *City and Territory in the Globalization Age*; Editorial Universitat Politècnica de València: Valencia, Spain, 2018; pp. 709–718.

36.   Steinberg, F. Conservation and rehabilitation of urban heritage in developing countries. *Habitat Int.* **1996**, *20*, 463–475. [CrossRef]

37.   Angl, J. *Historic Cairo—A Plea for World Heritage in Danger*; World Heritage Watch e.V.: Berlin, Germany, 2018.

38.   Bianca, S. *Urban form in the Arab World: Past and Present*; Thames & Hudson: London, UK, 2000.

39.   JCP. *Study on Conservation and Redevelopment of Historical Center of Baghdad City*; Amanat al Asisema: Baghdad, Iraq, 1984.

40.   Aboukhater, R. Culture-led urban development initiative in a world heritage city: The case study of the old city of Damascus in Syria. *Int. Arch. Photogramm. Remote Sens. Spat. Inf. Sci.* **2020**, *2020*, 573–580. [CrossRef]

41.   Ragab, T.S. The crisis of cultural identity in rehabilitating historic Beirut-downtown. *Cities* **2011**, *28*, 107–114. [CrossRef]

42.   The Municipality of Baghdad, Cartographer. *Land Uses Plane of Baghdad*; The Municipality of Baghdad: Baghdad, Iraq, 2016.

43.   Rashed, H. *Sustainable Urban Development in Historic Cairo*; University of Nottingham: Nottingham, UK, 2013.

44. The Municipality of Baghdad. *The Development of Al-Rasheed Street and the Tigris River Area Development Project*; The Municipality of Baghdad: Baghdad, Iraq, 2009.
45. Al-Hasani, M. Urban space transformation in old city of Baghdad--integration and management. *Megaron Archit.* **2012**, *7*, 79.
46. Al-Silq, G. Baghdad-Image and Memories. In *DC PAPERS: Revista de Crítica y Teoría de la Arquitectura*; Escola Tècnica Superior d'Arquitectura de Barcelona: Barcelona, Spain, 2008; pp. 46–65.
47. The Municipality of Baghdad. *The Development of Al-Rasheed Street and the Area between Al-Rashid Street and the Tigris Riverfront*; The Municipality of Baghdad: Baghdad, Iraq, 2011.
48. Al-Akkam, A. Towards Environmentally Sustainable Urban Regeneration: A Framework for Baghdad City Centre. *J. Sustain. Dev.* **2012**, *5*, 58. [CrossRef]
49. Hiu, Y.; Hon, C. Critical social sustainability factors in urban conservation: The case of the central police station compound in Hong Kong. *Facilities* **2012**, *30*, 396–416.
50. Jokilehto, J. *A History of Architectural Conservation*; Butterworth-Heinemann: Oxford, UK, 1999.
51. EH. *Heritage Counts 2004: The State of England's Historic Environment*; Historic England: London, UK, 2004.

Disclaimer/Publisher's Note: The statements, opinions and data contained in all publications are solely those of the individual author(s) and contributor(s) and not of MDPI and/or the editor(s). MDPI and/or the editor(s) disclaim responsibility for any injury to people or property resulting from any ideas, methods, instructions or products referred to in the content.

Article

Joint Management Plans in World Heritage serial nominations: the case of Álvaro Siza's Modern Contextualism Legacy

Teresa Cunha Ferreira [1,*] , **Pedro Murilo Freitas** [1] , **Tiago Trindade Cruz** [1] and **Hugo Mendonça** [2]

[1] Centre for Studies in Architecture and Urbanism, Faculty of Architecture, University of Porto, 4150-564 Porto, Portugal; pfreitas@arq.up.pt (P.M.F.); tcruz@arq.up.pt (T.T.C.)

[2] Faculty of Architecture, University of Porto, 4150-564 Porto, Portugal

* Correspondence: tferreira@arq.up.pt

Abstract: One of the most important challenges faced by any listed cultural heritage is the development of a management system that conveys a resilient and integrated approach that can sustain its values for future generations. Management is one of the main factors affecting World Heritage Sites; thus, the increased complexity of a serial nomination enhances this risk. By integrating different stakeholders, a Joint Management Plan (JMP) is a key tool to settle common procedures and help different managers maintain a satisfactory balance in safeguarding the Outstanding Universal Value (OUV) in each component part. This paper aims to provide a framework for the development of JMPs for serial nominations, with support on the nomination proposal "Álvaro Siza's Architecture: A Modern Contextualism Legacy". Methods result from the cross-analysis of (i) policy analysis; (ii) archival research and digital documentation; (iii) collaborative strategies (surveys, interviews, workshops, meetings, consultations); (iv) fieldwork. Results confirmed that the development of JMPs must be sustained by an open and dynamic process, where engagement, mediation of conflicts, and flexibility are key principles. This work approaches a significant subject concerning the management of World Heritage serial nominations, focusing on JMPs for serial nominations, which are a rising trend in heritage management. A demonstration is applied to the WH nomination of works by Álvaro Siza, a prominent figure in worldwide contemporary architecture.

Keywords: World Heritage; serial nominations; Joint Management Plan; 20th-century architecture; modern heritage; Álvaro Siza

Citation: Ferreira, T.C.; Freitas, P.M.; Cruz, T.T.; Mendonça, H. Joint Management Plans in World Heritage serial nominations: the case of Álvaro Siza's Modern Contextualism Legacy. *Architecture* **2024**, *4*, 820–834. https://doi.org/10.3390/architecture4040043

Academic Editor: Johnathan Djabarouti

Received: 7 May 2024
Revised: 18 September 2024
Accepted: 29 September 2024
Published: 1 October 2024

Copyright: © 2024 by the authors. Licensee MDPI, Basel, Switzerland. This article is an open access article distributed under the terms and conditions of the Creative Commons Attribution (CC BY) license (https://creativecommons.org/licenses/by/4.0/).

1. Introduction

1.1. Joint Management Plans in World Heritage (WH) Serial Nominations

According to The Operational Guidelines for the Implementation of the World Heritage Convention [1], "Protection and Management" is a mandatory pillar for the nomination of properties to the WH List. While planned during preparation, management endorses and provides endurance to the whole process, setting "a realistic vision for the medium to long-term future of the property, including the changes and challenges that could arise from inscription" [2] (p. 89). As management is also considered the main factor affecting WH sites [3], the absence of a thoughtful and reliable management system in a proposal may undermine property inscription. Thus, the development of a management system that conveys a resilient and integrated approach that can sustain its values for future generations, as affirmed by several recent UNESCO Guidelines, remains one of the main challenges of WH nomination preparation [4].

The UNESCO Resource Manual Series on Managing Cultural World Heritage [5] recommends a management system integrating three elements: (i) the legal framework; (ii) the institutional framework; and (iii) resources. The recent literature [6] also points out the relevance of monitoring and evaluating the management framework of WH properties, guaranteeing that the management process is functioning as intended, according to established rules, and meeting external reporting requirements [7]. Therefore, the monitoring process will evaluate the implementation and impact of each Management Plan, and, if needed, a Heritage Impact Assessment (HIA) should also be implemented [8].

In turn, serial nominations are becoming a growing trend in WH properties by integrating several component parts in the same Outstanding Universal Value (OUV) framework while responding jointly to criteria, authenticity and integrity, and protection and management. By definition, serial properties are properties that "include two or more component parts related by clearly defined links", which "should reflect cultural, social or functional links over time that provide, where relevant, landscape, ecological, evolutionary or habitat connectivity" and "contribute to the Outstanding Universal Value of the nominated property as a whole in a substantial, scientific, readily defined and discernible way, and may include, inter alia, intangible attributes" [1] (par. 137). In total, by 2021, about 60% of all properties inscribed had fallen into the serial category [9], and this proportion might have increased in 2023.

After inquiring about the WH List for "serial property" on the WH List website [10], results revealed 154 entries, including 118 cultural properties (76.7%), 34 natural properties (22.1%), and 2 mixed properties (1.2%). Although it is not the intent of this article to thoroughly analyze the WH List, as a sample, the distribution by year of inscribed properties that have the term in their description after 2000 (Figure 1) is consistent with the idea that overarching serial nominations were developed, at a first stage, to contribute for natural properties inscription, recovering an instrument that already existed in the WH Convention since 1980 to integrate geographically distant areas with an expressive common OUV [11]. In the second stage, this strategy has evolved in recent decades as a procedure for enabling cultural properties to respond to the Global Strategy by developing new required thematic frameworks [12], showcasing unique cultural and mixed contributions for the credibility, balance, and representativity of the WH List [13]. Hence, the inquiry showed an exponential increase, with 30 properties (19.5% of entries) mentioning the term only in the last 5 years.

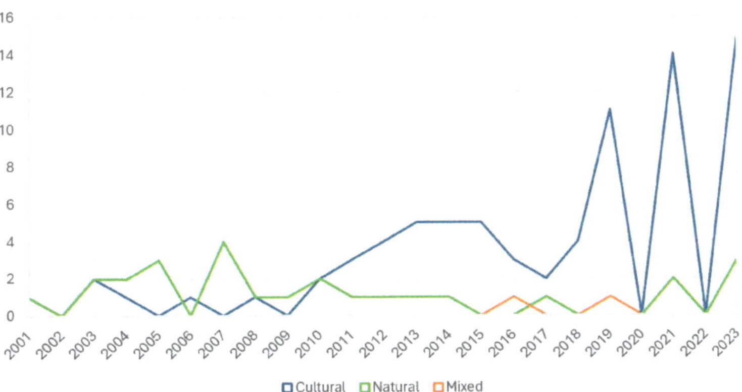

Figure 1. Properties mentioning the term "serial property" in the WH List by year (2001–2023).

This phenomenon of growing cultural serial nominations is creating new issues in developing management plans for serial WH properties. One topic of concern is the effect of a "collection of assets", hiding the number and extension of component parts inside a property [11]. Thus, the term "joint", referencing the idea of mutual or shared, although maintaining individual characteristics of the component parts of a given thematic framework, is emerging to define stronger management procedures and carefully define the OUV, the statements of authenticity and integrity, and the collaborative system for managing properties with growing complexity.

Hence, as a strategic document, a Joint Management Plan (JMP) is an essential tool providing an overarching perspective for the preservation of the physical and intangible substance of a WH property and the coordination of different stakeholders. Also, it enables the joint organization of individual Management Plans for each component part of a serial nomination, making its operational implementation more feasible. The JMP, in serial WH properties, must include different stakeholders, such as communities and organizations at different levels (local, regional, national, or even international), and settle common strategies for preserving the OUV and managing each component part.

However, there are still no specific guidelines for the development of JMPs, and very few examples of this kind of document have been applied to serial (national or transnational) nominations. Alongside the "Operational Guidelines" [1] updated biannually and UNESCO Resource Manuals publications [2,5], previous nominations are very resourceful for management preparation, providing examples, methods, and frameworks that can be potentially adapted to other properties. Despite that, in them, a "joint" approach still varies substantially.

Recently selected management plans, such as the Management Plan for the "Great Spa Towns of Europe" serial property (2021) [14], adopt an "overall management system", seeking to integrate different levels of protection as a priority. In other cases, such as the Management Plan for "The works of Jože Plečnik in Ljubljana—Human Centred Urban Design" (2021), a brief governance model is more focused on creating a joint articulation of conservation works since component parts were already protected by the same institutional body [15]. The Slovenian nomination innovated in the thematic framework of 20th-century architecture, as previous serial nominations of the works of Le Corbusier ("The Architectural Work of Le Corbusier, an Outstanding Contribution to the Modern Movement", 2016) or Frank Lloyd Wright ("The 20th-Century Architecture of Frank Lloyd Wright", 2019) did not present overarching management procedures or structured joint documents [16,17]. A more recent Management Plan for the serial nomination of the "Viking-Age Ring Fortresses" (2023) [18] aims for "integration, retention of value, sustainability and identification", closer to a more holistic definition. Also, despite the discussion at the International Expert Meeting on World Heritage and Serial Properties and Nominations in Ittingen, Switzerland, in 2010 [19], there is still the limited literature with in-depth methodologic and operational guidelines for drafting serial nominations. In this context, the existing literature continues to be scarce, insufficient, and even ambiguous, making the preparation process arduous and vulnerable to obstacles and backlashes in the nomination process.

This research improves current concepts of developing JMPs by integrating different experiences tested in the framework of the Burra Charter (created in 1979 and revised in 2013) and the UNESCO HUL approach, aiming to strengthen their application in the context of World Heritage serial nomination preparation. Namely, this represents the introduction of broader participatory techniques for the assessment of significance in the context of OUV definition while also integrating risk preparedness and adequate policy design for managing inscribed properties with complex component parts. With this framework and with the support of the specific case of the 20th-century serial WH nomination "Álvaro Siza's Architecture: A Modern Contextualism Legacy" [20], the research delves into a significant subject concerning the management of WH serial nominations, centered on the impact of Álvaro Siza and shedding light on a rising trend in heritage management. Hence, by adopting a holistic and inclusive approach, this article intends to provide a framework for effective and operational guidance for the management planning of serial nominations.

1.2. Álvaro Siza's Architecture: A Modern Contextualism Legacy

The "Álvaro Siza's Architecture: A Modern Contextualism Legacy" is a serial nomination for inscription on the World Heritage List submitted by Portugal in 2024. The property expresses the outstanding architecture by Álvaro Siza across the second half of the 20th century, which testifies to the critical revision of the Modern Movement principles toward a more contextual and humanist approach. The component parts (Figure 2) emerge as a result of the architectural development in the second half of the 20th century, addressing the specific conditions of the local contexts and producing alternative responses to the prevailing axioms of international Modernism while also contributing to the Post-Modernism debate.

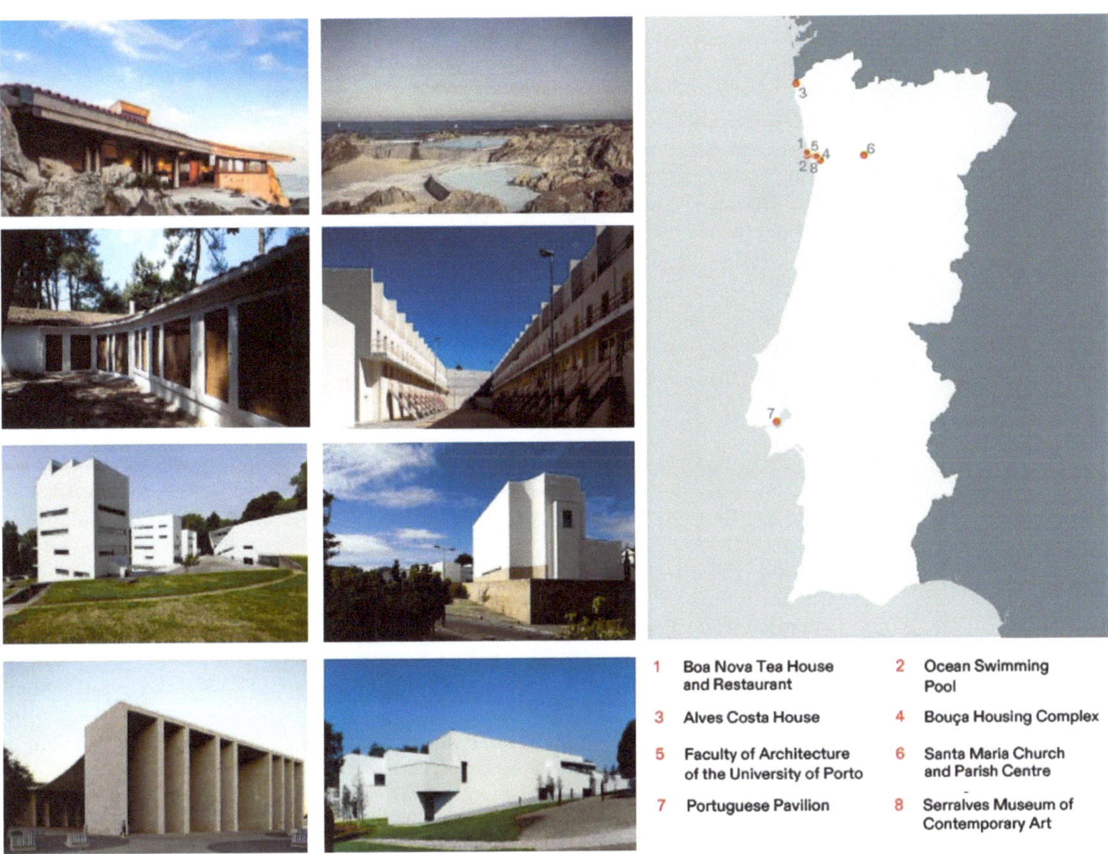

Figure 2. Selected component parts of the serial nomination "Álvaro Siza's Architecture: A Modern Contextualism Legacy" (2024).

Álvaro Siza is a prominent figure in contemporary architecture with more than 500 projects and almost 200 built works in 16 countries and on 4 continents. With over 100 prizes and distinctions, Siza holds 19 honorary doctorates and hundreds of dedicated publications.

The property includes a series of eight component parts located in Portugal, designed and built across the second half of the 20th century: the Boa Nova Tea House and Restaurant (1958); the Ocean Swimming Pool (1961); the Alves Costa House (1964); the Bouça Housing Complex (1975); the Faculty of Architecture of the University of Porto (1984); the Santa Maria Church and Parish Centre (1990); the Serralves Museum of Contemporary Art (1991); and the Portuguese Pavilion (1995). The component parts of this series comprise a variety of public-use buildings, social housing, a private house, museum and exhibition areas, university buildings, a teahouse and restaurant, a swimming pool, and a church.

2. Materials and Methods

2.1. Supporting Studies

For each aforementioned component part, a management plan was produced between 2022 and 2023 by an interdisciplinary team assessed by the UNESCO Chair "Heritage, Cities and Landscapes. Sustainable Management, Conservation, Planning, and Design", hosted at the Faculty of Architecture of the University of Porto. Also, the following supporting studies contributed to research and development:

2.1.1. Siza ATLAS: Filling the Gaps for World Heritage

This research project funded by the Foundation for Science and Technology (2021–2024) proposes to develop a comprehensive inventory of all of Álvaro Siza's built works and conduct a detailed analysis and documentation of 18 component parts selected for the World Heritage Tentative List. The Siza ATLAS research methodology is supported by the cross-analysis of different methods and tools:

1. Archival and bibliographic research;
2. Photogrammetric survey;
3. Virtual tours, 360 captions, and photographic survey;
4. Three-dimensional drawings of constructive sections and details.

All of these methods and tools are supported by direct confrontation and analysis in situ of the built works.

2.1.2. Keeping It Modern: Ocean Swimming Pool

This research project funded by the Getty Foundation (2020–2023) consists of developing research and best practices and preparing a Conservation and Management Plan for the Ocean Swimming Pool in Leça da Palmeira, designed by Álvaro Siza between 1960 and 1966. Awarded by the "Keeping It Modern" program, the project contributed to settling a framework for the management plans of the component parts of this nomination. Supported by the Burra Charter methodology [21], the Conservation and Management Plan of the Ocean Swimming Pool was based on a holistic and integrated approach as follows:

1. Understanding the place and documenting the history of design and construction;
2. Assessing the cultural significance;
3. Analyzing vulnerabilities and the physical condition of the building (including inspection and diagnosis and the conservation of reinforced concrete);
4. Developing policies, including maintenance and use;
5. Defining the future monitoring and implementation.

2.2. Participation Strategies

Due to the complexity and diversity of stakeholders, serial nominations require a collaborative and inclusive approach. This is achieved by adopting procedures of identification, community engagement, and participation strategies. Their interaction remains a key issue during the planning process. Several methods and tools (surveys, interviews, workshops, consultations, meetings) were implemented to define the following:

1. Attributes and values (of the OUV);
2. Vulnerabilities and factors affecting the property;
3. Strategies and actions for the future management of change (Figure 3).

Since 2021, the required section Stakeholders and Participation in the "Operational Guidelines" [1] (para. 123) set up new community-driven approaches for nomination preparation. Thus, participation strategies tested in a similar context in the aforementioned project "Keeping It Modern: Ocean Swimming Pool" [22] were selected through a corresponding identification and cross-examination of the different stakeholders involved. Through these strategies, sufficient data were combined for the definition of management and conservation planning. This was achieved by necessarily combining further recommendations [23,24], including the UNESCO HUL approach [25], and expanding the value-based (or "value-led") planning as suggested by The UNESCO Resource Manual [5].

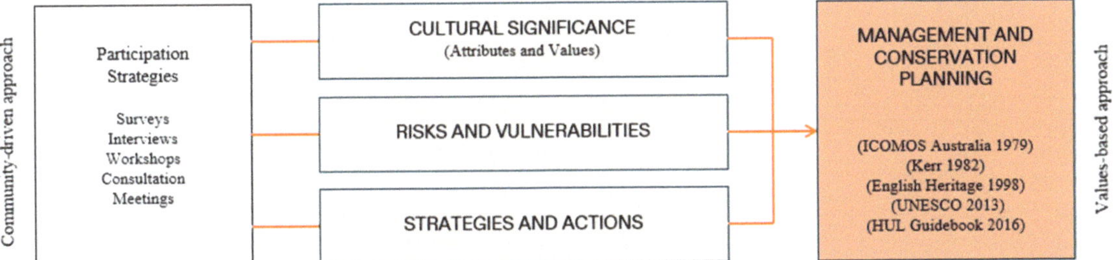

Figure 3. Management and Conservation Planning for the serial nomination "Álvaro Siza's Architecture: A Modern Contextualism Legacy" (2024).

2.3. Data Collection

The methodology for data collection was supported by the cross-analysis of distinct methods and tools:

1. Policy analysis (serial nominations, management plans, UNESCO Guidelines, applicable legislation);
2. Archival research and digital documentation;
3. Participative strategies (surveys, interviews, workshops, meetings, consultations);
4. Fieldwork (data collection, surveys, and in-depth analysis of each component part).

Participation techniques were one of the most innovative issues in the methodologic approach of the JMP and individual MPs, namely, in the assessment of cultural significance (attributes and values), analysis of vulnerabilities, and the definition of management strategies and actions.

3. Results

The Results section intends to respond to the research question, "How to develop collaborative Joint Management Planning for serial nominations?" With this purpose, this section develops on the different sections of the JMP, providing exemplifications of the specific case study of "Álvaro Siza's Architecture: A Modern Contextualism Legacy".

Complying with the three pillars of the OUV as defined by UNESCO—1. Criteria; 2. Authenticity and Integrity; and 3. Protection and Management [2]— the JMP's structure is supported by the definition of the cultural significance of the property (Chapter 1) and the development of a Joint Management System based on legal compliance, institutions, and resources that may sustain its future (Chapter 2). As a result of this fundamental value-based approach to conservation planning, the vulnerabilities affecting the property (Chapter 3) are then thoroughly analyzed, as are the base to set up strategies and actions for the careful management of change (Chapter 4) and their subsequent monitoring and implementation (Chapter 5).

Table 1 shows the JMP structure developed for "Álvaro Siza's Architecture: A Modern Contextualism Legacy" serial WH nomination:

Table 1. JMP Structure.

Chapter	Description
1. Registration Process for the World Heritage List	This chapter summarizes how the property complies with criteria (in relation to the attributes of the OUV), as well as with authenticity and integrity. General descriptions are followed by a synthesis of how each component part expresses the attributes, authenticity, and integrity.
2. Joint Management System	This chapter defines how the property complies with protection and management requirements by detailing the Joint Management System proposed for the management of the serial property, namely, 1. legal framework, 2. institutional framework, and 3. resources.
3. State of Conservation and Factors Affecting the Property	This chapter addresses the conservation conditions of the component parts and defines the factors affecting the property, followed by specific insights on each component part.
4. Strategies and Action Plan	This chapter defines the strategies for the future management of change, structured in 1. strategic objectives and 2. action lines to be further detailed in specific 3. actions for the achievement of defined 4. outputs.
5. Monitoring, Impact Assessment, and Implementation	This chapter describes the monitoring activities and the process of periodic reporting and heritage impact assessments, as well as a phasing implementation proposal (detailing priority actions on yearly basis and availability of resources).

3.1. Registration Process for the World Heritage List

This chapter develops on the justification of the OUV and, more specifically, the gap being filled to contribute to the representativity, balance, and credibility of the WH List. Hence, it is focused on the first two pillars of the OUV, namely, 1. Criteria; 2. Authenticity and Integrity, as defined in The Operational Guidelines for the Implementation of the World Heritage Convention [1]. In a value-based approach, cultural significance is the guiding force for future joint decision-making regarding their preservation.

3.1.1. Criteria and Attributes

This item relates to the first pillar of the OUV, namely, the criteria selected for the inscription of the property on the WH List from the six criteria available for cultural sites, in articulation with the requirements of authenticity and integrity, as well as the protection and management according to The Operational Guidelines for the Implementation of the World Heritage Convention [1].

Table 2 demonstrates how criteria and attributes are related in "Álvaro Siza's Architecture: A Modern Contextualism Legacy".

Table 2. Relation between Criteria and Attributes.

Criteria	Attributes (OUV—Siza WH)
Criterion (ii): To exhibit an important interchange of human values over a span of time or within a cultural area of the world on developments in architecture or technology, monumental arts, town-planning, or landscape design	Attribute 1: Architecture responsive to a physical, social, and historical context
	Attribute 2: Integration of international and local references
Criterion (iv): To be an outstanding example of a type of building, architectural or technological ensemble, or landscape, which illustrates (a) significant stage(s) in human history	Attribute 3: Sculptural volumetric expression
	Attribute 4: Oriented spatial experiences
	Attribute 5: Total work of art, within details, furniture, and artworks.

3.1.2. Authenticity and Integrity

This item relates to the second pillar of the OUV, responding to the authenticity and integrity requirements as defined in The Operational Guidelines for the Implementation of the World Heritage Convention [1]. Regarding Alvaro Siza's nomination proposal, these criteria have been demonstrated by his involvement in recent conservation works, fully respecting the original design and making only small updates related to current comfort or regulation standards:

Authenticity. This item must comply with the requirements of authenticity, as defined in the Nara Document on Authenticity [26]: (i) form and design; (ii) materials and substance; (iii) use and function; (iv) location and setting; (v) spirit and feeling; (vi) traditions, techniques, and management systems. Compliance must be expressed globally and for each component part, not having suffered significant changes, and maintain the general authenticity of the original design.

Regarding this case study, minor changes have been carried out to adapt to the current living standards and legislation in compliance with the preservation of authenticity. All conservation works have been carried out with the best methodologies to preserve their authenticity, benefiting from the supervision of heritage safeguarding bodies for listed buildings with full respect for the attributes of the OUV. All the measures necessary for the future preservation of their cultural meaning are individually defined in the Management Plans produced for each component part;

Integrity. This item aims to demonstrate that the property is a testimony of integrity (both as a whole and in the individual components) assessed by the following vectors presented in The Operational Guidelines for the Implementation of the World Heritage Convention [1]: (i) includes all elements necessary to express its OUV; (ii) is of adequate size to ensure the complete representation of the features and processes which convey the property's significance; (iii) suffers from adverse effects of development and/or neglect.

Regarding this case study, each Buffer Zone is of adequate size to include critical elements of its setting, as further described in detail for each of them. Even though the landscape has changed in some cases, the immediate setting remains largely intact, while the views from the building toward its surrounding landscape are protected.

3.2. Joint Management System

This chapter relates to the third pillar of the OUV (Protection and Management) and is an essential tool to preserve the OUV and ensure the collaborative management of the property. This is the core of the JMP.

The UNESCO Resource Manual Series on Managing Cultural World Heritage [5] recommends a management system integrating three elements:

1. The Legal framework, a set of national decrees, laws, and other norms aligned with international recommendations aiming to "provide sufficient legal and regulatory tools for the protection of cultural heritage" [5] (p. 65);
2. The Institutional framework should be flexible enough and "provide for efficient decision-making and facilitate all processes of the management system" [5] (p. 71);
3. Resources, "inputs", which "fall into three broad categories—human, financial and intellectual"—and "make a management system operate to conserve and manage cultural heritage" [5] (p. 73).

An important issue in regulating the Joint Management System is establishing a close correlation between these three categories, assuming protection mechanisms in force, combining current institutions and their responsibilities, and providing a balance between stakeholders (including communities). This operation must consider long-term challenges and have an adequate source of financial resources and staff. Also, in a serial nomination, the Joint Management System requires multiple stakeholders' engagement (Government organizations, NGOs, Municipal Councils, Owners, Communities, and rights holders), and the joint management body may assume distinct formal concretizations, according to the legislation of each country and the specificities of each property.

In the case of "Álvaro Siza's Architecture: A Modern Contextualism Legacy", the creation of the "Álvaro Siza World Heritage Association" is proposed by presenting the respective statutes in the JMP as the most suitable format for an overarching body engaging multiple stakeholders. The Association will act as a cooperative platform for future management, including existing ones, while also stimulating knowledge sharing and cooperation between managers, owners, and institutions. The Association, thus, becomes the formal overarching instance for responding to UNESCO periodic reporting requirements, among other tasks such as dissemination and monitoring, through its respective focal point, the Faculty of Architecture of the University of Porto (Figure 4).

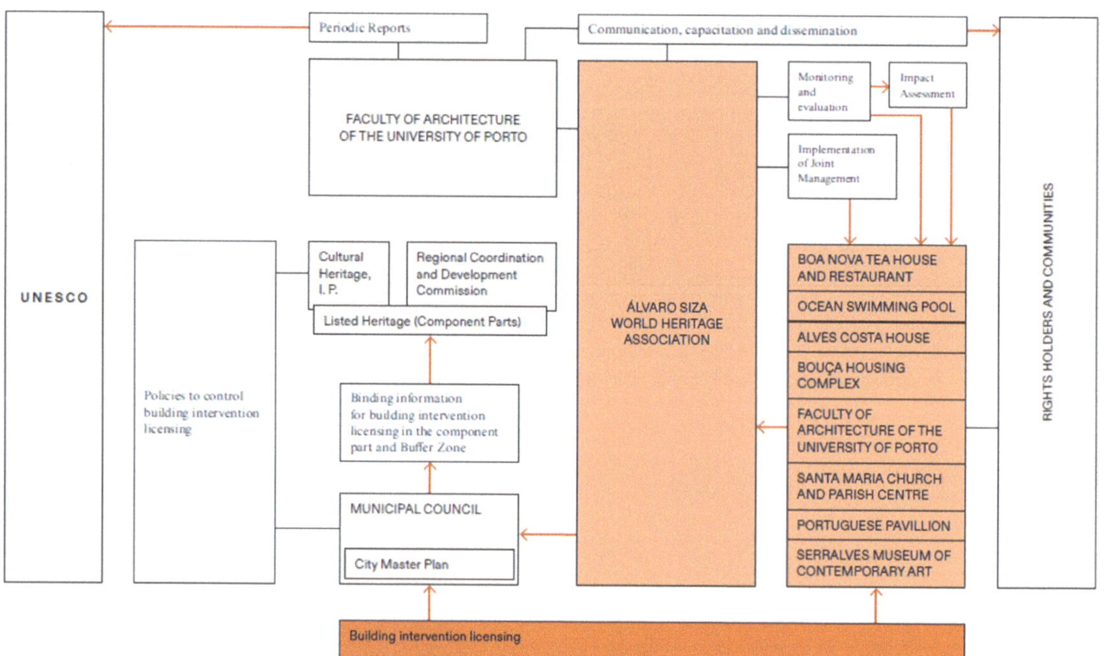

Figure 4. Joint Management System for the "Álvaro Siza's Architecture: A Modern Contextualism Legacy" (2024).

The Association is a key tool for the Joint Management System, as well as for the engagement, participation, and empowerment of all the stakeholders. This is in line with the participation approach defined in the methodology for the JMP and individual Management Plans, endorsing an innovative approach responding to the most recent guidelines from UNESCO and ICOMOS.

3.3. State of Conservation and Factors Affecting the Property

This chapter addresses the condition of the property through its component parts, including the identification of their state of conservation (and possible causes of degradation). This is closely interrelated with the main factors affecting the property, classified according to UNESCO [27].

3.3.1. State of Conservation

The state of conservation must be sustained on a regular review within a framework of monitoring processes for WH properties, as specified within The Operational Guidelines for the Implementation of the World Heritage Convention [1]. State of conservation includes material conditions as a principal factor but may also be indicative of different vulnerabilities that may not be evident. Also, in a joint approach, management must be aware of imbalances in the conditions of the component parts through an integrated perspective and how breadth differences may affect the series.

Regarding "Álvaro Siza's Architecture: A Modern Contextualism Legacy", the state of conservation of the component parts is generally very good. Over the past ten years, a series of conservation works were carried out, coordinated by Álvaro Siza with the supervision of the Cultural Heritage, I. P. (for listed buildings, former General Directorate of Cultural Heritage—DGPC and the competent regional bodies). Table 3 shows a summary of these interventions.

Table 3. State of Conservation of the Property.

Component Part	Recent Intervention	Architect	Promoter	Condition
Boa Nova Tea House and Restaurant	2013–2014	Álvaro Siza	Municipal Council of Matosinhos	Very Good
Ocean Swimming Pool	2018–2021	Álvaro Siza	Municipal Council of Matosinhos	Very Good
Alves Costa House	Periodic Maintenance	-	Alexandre Alves Costa	Very Good
Bouça Housing Complex	2000–2006	Álvaro Siza	Águas Férreas Condominium	Good
Faculty of Architecture of the University of Porto	2016–2017	Álvaro Siza	University of Porto	Very Good
Santa Maria Church and Parish Centre	2022–present	Álvaro Siza	Municipal Council of Marco de Canaveses	Very Good
Portuguese Pavilion	2016–present	Álvaro Siza	University of Lisbon	Very Good
Serralves Museum of Contemporary Art	Periodic Maintenance	Álvaro Siza	Serralves Foundation	Very Good

3.3.2. Factors Affecting the Property

The factors affecting the property are provided by UNESCO in a list that consists of a series of 14 primary factors, each encompassing a number of secondary factors [27]. The function of a joint approach is to examine the most common factors despite geographical distances, context, and, thus, vulnerabilities. This analysis may enhance the development of risk assessments and provide immediate response by the entity responsible for the Joint Management System.

In the case of "Álvaro Siza's Architecture: A Modern Contextualism Legacy", the factors are structured according to the three classes of factors defined in The Operational Guidelines for the Implementation of the World Heritage Convention [1], such as all components being located in urban development areas or infrastructure being present (Factor Class 1). Also, weathering, heavy rainfall, and fire are common threats to the location of a few components in the proximity of waterfronts (sea or river), leading to some factors, such as coastal erosion and rising sea levels (Factor Class 2). Finally, misuse and vandalism are among the most prominent relating to Factor Class 3.

Table 4 provides a summary of each factor affecting the component parts and, thus, the property as a whole.

Table 4. Summary of all factors affecting the property.

Factor Class	Factor Type	Component Parts *							
		1	**2**	**3**	**4**	**5**	**6**	**7**	**8**
1. Development Pressures and Management Response	Real Estate Development	•			•		•		
	Infraestructural Changes					•			
	Accessibility			•		•	•		
2. Environmental Pressures, Natural Hazards, and Risk Preparedness	Weathering	•	•	•	•	•	•	•	•
	Atmosferic CO$_2$ Concentration	•	•		•	•	•	•	•
	Chloride Action	•	•			•		•	
	Sea Level Rise		•					•	
	Coastal Erosion	•	•						
	Heavy Rainfall	•		•	•	•	•	•	•
	Strong Winds	•		•					
	Fire	•	•	•	•	•	•	•	•
	Storm Surge	•	•						
	Earthquake/Tsunami							•	
3. Visitation, Other Human Activities, and Sustainable Use	Vandalism	•	•		•	•	•	•	
	Misuse	•	•			•	•	•	•
	Lack of Sustainable Visitor Engagement			•	•				

* (1) Boa Nova Tea House and Restaurant; (2) Ocean Swimming Pool; (3) Alves Costa House; (4) Bouça Housing Complex; (5) Faculty of Architecture of the University of Porto; (6) Santa Maria Church and Parish Centre; (7) Portuguese Pavilion; (8) Serralves Museum of Contemporary Art.

3.4. Strategies and Action Plan

This chapter defines the strategies for the preservation and enhancement of the OUV while providing measures to mitigate the impacts of the factors affecting the property and defining actions for adequate management of change. The multiple stakeholders (especially managers) in a serial property may cause a lack of communication and purpose toward decision-making. Hence, actions must be framed precisely and closely defined by resource availability, time, and responsibility.

Regarding "Álvaro Siza's Architecture: A Modern Contextualism Legacy", the following Strategies and Strategic Objectives are defined in the JMP and further complemented in each component part's Management Plans:

1. Overarching Strategies, focused on (A) Management Integration and Shared Knowledge Network;
2. Operational Strategies, which include (B) Spatial Planning, Risk Management, and Climate Change Adaptation; (C) Interpretation, Communication, and Community Engagement; and (D) Conservation, Maintenance, and Appropriate Use.

3.4.1. Strategies

Overarching Strategies. They are the ones that gather all the fundamental actions for the World Heritage Management System implementation. They are meant to be the primary source of guidance for managing the component parts' daily challenges to preserve the OUV;

Operational Strategies. They are a broader set that combines operative actions for the preservation of each component part's OUV. It ranges from compliance with national legislation and international guidelines to the implementation of the best conservation practices and adequate use of the component parts while enhancing and disseminating their OUV to a broader public.

3.4.2. Strategic Objectives

(A) Management Integration and Shared Knowledge Network. The first strategic objective aims to guarantee the integration of all managers of the component parts. Its implementation is based on the creation of a common platform for sharing experiences, information, and documentation on which property managers can rely for bridging difficulties and making decisions based on institutional consensus;

(B) Spatial Planning, Risk Management, and Climate Change Adaptation. The second strategic objective aims to enforce the adoption of the current legal framework to safeguard the component parts' OUV. It also establishes mandatory measures for managing the integrity of each component part's surroundings and their relationship with the specific context, including the adaptations foreseen by climate change;

(C) Interpretation, Communication, and Community Engagement. The third strategic objective aims to qualify a framework through which the cultural significance of each component part is interpreted and updated as WH. This includes actively involving the community in sustaining the component parts' OUV and developing an efficient communication system based on compromise with community needs.

(D) Conservation, Maintenance, and Appropriate Use. The fourth strategic objective aims to ensure the preservation of the component part's original design and cultural significance while adapting it to current and future needs as a means for conserving efficiency and sustainability through time. It has the purpose of providing guiding principles for periodic maintenance actions in line with the best conservation practices.

Table 5 demonstrates how Strategies and Strategic Objectives were able to organize Action Lines proposed in the JMP of "Álvaro Siza's Architecture: A Modern Contextualism Legacy" serial nomination.

Table 5. Hierarchy of Strategies, Strategic Objectives, and Action Lines.

Strategies	Strategic Objectives	Action Lines
Overarching Strategies	A. Management Integration and Shared Knowledge Network.	A.1. Stakeholders
		A.2. Research and Knowledge
		A.3. Sustainable Development Goals
Operational Strategies	B. Spatial Planning, Risk Management, and Climate Change Adaptation	B.1. Spatial Planning
		B.2. Landscape Management
		B.3. Buffer Zone Preservation
		B.4. Risk Management
		B.5. Climate Change Adaptation
	C. Interpretation, Communication, and Community Engagement	C.1. Interpretation
		C.2. Communication
		C.3. Community Engagement
	D. Conservation, Maintenance, and Appropriate Use	D.1. Maintenance and Archive Database
		D.2. Housekeeping
		D.3. Accessibility
		D.4. Use and Occupancy
		D.5. Security and Fire Prevention

3.5. Monitoring, Impact Assessment, and Implementation

Based on the previously presented management strategies, this chapter has the objective of providing a comprehensive timeframe for defining priorities, as well as a hierarchy for the implementation of conservative actions, respecting funding and other resources available.

The development of this process includes the perception of the management system as a whole (First level of monitoring), as well as the attention to threats and impacts on OUV and the characteristics of each component part (second level of monitoring). Impact Assessment may be a resourceful instrument, as well as establishing priorities and an implementation program in case of need.

Table 6 demonstrates how the first level of monitoring is proposed in the JMP of "Álvaro Siza's Architecture: A Modern Contextualism Legacy" serial nomination.

Table 6. Thematic Framework (First Level of Monitoring).

Subject	Indicators	Periodicity	Responsible
OUV	Verification that the Outstanding Universal Value of the property and its component parts is intact.	Annually *	State Party
Authenticity and Integrity	Evaluation of the component parts' condition under ranges of authenticity and integrity.	Anually	Cultural Heritage, I.P.
Conservation Status	Monitoring the physical condition of the property through regular inspections and extensive documentation of anomalies and interventions.	Anually	Manager
Finance	Annual budget for maintenance, repair, and conservation.	Anually	Owner
Management	Evaluation of Management Plans.	Anually	Álvaro Siza World Heritage Association
Buffer Zones	Monitoring irregular interventions inside the Buffer Zones and impact evaluation on the Outstanding Universal Value.	Anually	Cultural Heritage, I. P., Regional Coordination, and Development Commission

* Besides annual periodicity of monitoring the Outstanding Universal Value, authenticity, and integrity of the property, every 6 years, the State Party must also submit a Periodic Report, according to article 29 of the WH Convention.

4. Conclusions

This paper does not intend to establish a univocal or exhaustive model of JMP but rather to leave some guidelines and perspectives that can contribute to the future development of WH serial nominations. It is not possible to establish rigid and static models of management models; they must be dynamic processes and adapted to the specificity of each case.

With regard to serial nominations, the collaborative and inclusive approach of all stakeholders is a determining factor throughout the process. For this purpose, the nomination proposal for "Álvaro Siza's Architecture: A Modern Contextualism Legacy" integrated different collaborative strategies (surveys, interviews, workshops, meetings, consultations), involving all stakeholders from the initial phase of defining attributes and creating the management model ("Álvaro Siza World Heritage Association").

One innovative aspect of this research is the participation approach included in the methodology for the development of JMPs and individual MPs (surveys, interviews, workshops, meetings, and consultations), engaging a wide range of stakeholders and, thus, potentially impacting society, influencing perceptions of heritage conservation, or enhancing community involvement in heritage sites.

In these processes, knowledge sharing and transparency, mediation of conflicts, and definition of consensus are fundamental. The limitations of this work are related to the time and resources available for the work. Due to the participation approach adopted in this process, there are biases and constraints related to the samples (in some cases, such as private houses, samples are reduced), as well as challenges in data collection during the COVID-19 period. However, this framework for the JMP is open and flexible to future developments and adjustments during its implementation. Hence, this paper's findings are open to future research, namely, on the implementation of this holistic, integrated, and participative approach to other heritage properties, not only WH sites but also to managing change in built heritage from a wide perspective.

Author Contributions: Conceptualization, T.C.F. and P.M.F.; methodology, T.C.F. and P.M.F.; investigation, T.C.F., P.M.F., T.T.C. and H.M.; writing—original draft preparation, T.C.F. and P.M.F.; supervision, T.C.F.; project administration, T.C.F.; funding acquisition, T.C.F. All authors have read and agreed to the published version of the manuscript.

Funding: This study is co-financed by the European Regional Development Fund (ERDF) through COMPETE 2020—Operational Programme for Competitiveness and Internationalisation (OPCI) and by national funds through FCT, under the scope of the POCI-01-0145-FEDER-007744 project, 2020.01980.CEECIND and BPI La Caixa.

Institutional Review Board Statement: Not applicable.

Informed Consent Statement: Informed consent was obtained from all subjects involved in this study regarding surveys and participatory activities.

Data Availability Statement: Data available on request due to privacy restrictions.

Acknowledgments: This research is framed within the UNESCO Chair *"Heritage, Cities, and Landscapes. Sustainable Management, Conservation, Planning and Design"*.

Conflicts of Interest: The authors declare no conflict of interest.

References

1. UNESCO. *Operational Guidelines for the Implementation of the World Heritage Convention (WHC.21/01 2021)*; UNESCO: Paris, France, 2021; Available online: https://whc.unesco.org/document/190976 (accessed on 3 May 2024).
2. UNESCO. *Preparing World Heritage Nominations. Resource Manual*, 2nd ed.; UNESCO: Paris, France; ICCROM: Rome, Italy; ICOMOS: Charenton-le-Pont, France; IUCN: Gland, Switzerland, 2011; Available online: https://whc.unesco.org/en/preparing-world-heritage-nominations/ (accessed on 3 May 2024).
3. Veillon, R. *State of Conservation of World Heritage Properties—A Statistical Analysis (1979–2013)*; UNESCO: Paris, France, 2014; Available online: https://whc.unesco.org/en/documents/134872 (accessed on 3 May 2024).
4. Feilden, B.; Jokkilehto, J. *Management Guidelines for World Cultural Heritage Sites*; ICCROM: Rome, Italy, 1998; Available online: https://www.iccrom.org/publication/management-guidelines-world-cultural-heritage-sites (accessed on 3 May 2024).
5. UNESCO. *Managing Cultural World Heritage. Resource Manual*, 4th ed.; UNESCO: Paris, France; ICCROM: Rome, Italy; ICOMOS: Charenton-le-Pont, France; IUCN: Gland, Switzerland, 2013; Available online: https://whc.unesco.org/en/managing-cultural-world-heritage (accessed on 3 May 2024).
6. Breda Vásquez, I.; Conceição, P.; Brandão Alves, F.; Rocha, C.; Coimbra, I.; Silva, D.; Sousa, A.R.; Tavares, D. *Methodology for the Development of Management Plans for Urban World Heritage Sites*; FEUP: Porto, Portugal, 2020; Available online: http://www.atlaswh.eu/files/publications/20_1.pdf (accessed on 6 May 2024).
7. Boccardi, G.; Stovel, H. Monitoring World Heritage—Conclusions of the International Workshop. In *Monitoring World Heritage. World Heritage 2002: Shared Legacy, Common Responsibility, Associated Workshops, Vicenza, Italy, 11–12 November 2002*; UNESCO: Paris, France, 2004; Available online: https://unesdoc.unesco.org/ark:/48223/pf0000136571 (accessed on 6 May 2024).
8. Court, S.; Jo, E.; Mackay, R.; Murai, M.; Therivel, R. *Guidance and Toolkit for Impact Assessments in a World Heritage Context*; UNESCO: Paris, France; ICCROM: Rome, Italy; ICOMOS: Charenton-le-Pont, France; IUCN: Gland, Switzerland, 2022; Available online: https://whc.unesco.org/en/guidance-toolkit-impact-assessments (accessed on 6 May 2024).
9. Wolfling-Assa, O.; Alon-Mozes, T.; Liberty-Shalev, R. Serial Properties and Heritage Interpretation. Lessons from the Israeli Biblical Tels Inscription. *Int. J. Herit. Stud.* **2024**, *30*, 519–539. [CrossRef]
10. *World Heritage List*; UNESCO: Paris, France. Available online: https://whc.unesco.org/en/list/ (accessed on 6 May 2024).
11. Engels, B.; Ohnesorge, B.; Burmester, A. *Nominations and Management of Serial Natural World Heritage Properties. Present Situation, Challenges and Opportunities*; Federal Agency for Nature Conservation: Bonn, Germany, 2009; Available online: https://portals.iucn.org/library/sites/library/files/documents/Rep-2009-005.pdf (accessed on 6 May 2024).
12. UNESCO. *Expert Meeting on the "Global Strategy" and Thematic Studies for a Representative World Heritage List (UNESCO Headquarters, 20–22 June 1994)*; UNESCO: Paris, Italy, 1994; Available online: https://whc.unesco.org/archive/global94.htm (accessed on 6 May 2024).
13. Vileikis, O. Monitoring Serial Transnational World Heritage—The Central Asian Silk Roads Experience. *Hist. Environ. Policy Pract.* **2016**, *7*, 260–273. [CrossRef]

14. UNESCO. *Nomination of The Great Spas of Europe for inclusion on the World Heritage List. Volume III: Property Management Plan*; UNESCO: Paris, Italy, 2021; Available online: https://whc.unesco.org/document/177650 (accessed on 27 August 2024).
15. Museum of Architecture and Design. *Ljubliana: The Timeless, Human Capital Designed by Jože Plečnik. Nomination for Inscription on the World Heritage List*; MAO: Ljubliana, Slovenia, 2020; Available online: https://whc.unesco.org/document/181046 (accessed on 6 May 2024).
16. Fondation Le Corbusier. *The Architectural Work of Le Corbusier, an Outstanding Contribution to the Modern Movement. Nomination to the World Heritage List Presented by Germany, Argentina, Belgium, France, India, Japan and Switzerland*; FLC: Paris, France, 2016; Available online: https://whc.unesco.org/uploads/nominations/1321rev.pdf (accessed on 6 May 2024).
17. Frank Lloyd Wright Building Conservancy. *The 20th-Century Architecture of Frank Lloyd Wright. Nomination to the World Heritage List by the United States of America (2016) Revised 2019*; FLWBC: Chicago, IL, USA, 2019; Available online: https://whc.unesco.org/document/170692 (accessed on 6 May 2024).
18. Danish Agency for Culture and Palaces. *Nomination of Viking-Age Ring Fortresses for Inclusion on the World Heritage List*; DACP: Copenhagen, Danmark, 2023; Available online: https://whc.unesco.org/document/189404 (accessed on 27 August 2024).
19. Swiss Federal Office of Culture. Swiss Federal Office of Culture. UNESCO World Heritage: Serial Properties and Nominations. In Proceedings of the International Expert Meeting on World Heritage and Serial Properties and Nominations, Ittingen, Switzerland, 25–27 February 2010; SFOC: Bern, Switzerland, 2010. Available online: https://whc.unesco.org/document/124860 (accessed on 6 May 2024).
20. Faculty of Architecture of the University of Porto. *Álvaro Siza's Architecture: A Modern Contextualism Legacy. Nomination for Inscription in the World Heritage List*; FAUP-CEAU: Porto, Portugal, 2024; *submitted*.
21. ICOMOS Australia. *The Burra Charter: The Australia ICOMOS Charter for Places of Cultural Significance*; ICOMOS Australia: Burwood, Australia, 2013; Available online: https://australia.icomos.org/publications/burra-charter-practice-notes/ (accessed on 6 May 2024).
22. Cunha Ferreira, T.; Freitas, P.M.; Frigolett, C.; Mendonça, H.; Tarrafa Silva, A. The contribution of stakeholder engagement to cultural significance assessment: The case of values-based conservation management planning for the Ocean Swimming Pool, Portugal. *Built Herit.* **2024**, *8*, 1–18. [CrossRef]
23. Kerr, J.S. *Conservation Plan*, 7th ed.; ICOMOS Australia: Burwood, Australia, 2013; [1st. ed., 1982]; Available online: https://australia.icomos.org/publications/the-conservation-plan// (accessed on 30 September 2024).
24. English Heritage. *Conservation Plans in Action: Proceedings of the Oxford Conference*; Clark, K., Ed.; English Heritage: London, UK, 1999.
25. World Heritage Institute of Training and Research for the Asia and the Pacific Region. *The HUL Guidebook: Managing Heritage in Dynamic and Constantly Changing Urban Environments*; a practical guide to UNESCO's Recommendation of the Historic Urban Landscape; UNESCO-WHITRAP: Shanghai, China, 2016.
26. ICOMOS. *Nara Document on Authenticity*; ICOMOS: Nara, Japan, 1994; Available online: https://www.icomos.org/en/179-articles-en-francais/ressources/charters-and-standards/386-the-nara-document-on-authenticity-1994 (accessed on 7 May 2024).
27. *List of Factors Affecting the Properties*; UNESCO: Paris, France, 2024; Available online: https://whc.unesco.org/en/factors (accessed on 6 May 2024).

Disclaimer/Publisher's Note: The statements, opinions and data contained in all publications are solely those of the individual author(s) and contributor(s) and not of MDPI and/or the editor(s). MDPI and/or the editor(s) disclaim responsibility for any injury to people or property resulting from any ideas, methods, instructions or products referred to in the content.

Article

Building Home in Exile: The Role of Intangible Cultural Heritage, Crafts, and Material Culture Among Resettled Syrians in Liverpool, UK

Ataa Alsalloum 

School of Architecture, University of Liverpool, Liverpool L69 7ZN, UK; ataa2@liverpool.ac.uk;
Tel.: +44-(0)151-794-9596

Abstract: Since the onset of the Syrian conflict in 2011, millions of Syrians have sought refuge globally, with thousands resettling in the UK. Despite their displacement, Syrians have brought with them a rich array of inherited knowledge and traditions, collectively known as intangible cultural heritage (ICH). The construction of domestic spaces by these settlers and their struggle to feel at home have emerged as important topics in migration studies, particularly when housing issues are considered as a critical aspect of their transcultural social engagement and the evolving boundaries of their identity and belonging. However, the role of ICH, along with the related crafts and movable objects, in the home-making practices of forced migrants remains under-researched. This gap is especially significant given that the UK recently ratified the 2003 UNESCO Convention on the safeguarding of ICH after a decade-long delay. Through in-depth semi-structured interviews conducted in the interviewees' native Arabic within their home environments and supported by an observational study, this research explores how resettled Syrians in Liverpool integrate traditional ICH practices into their new homes, focusing on the dynamic relationship between the intangible and built heritage. By examining how intangible knowledge and movable objects interplay in creating a 'Syrian home', this study contributes to discussions on community engagement and the role of memory in conservation. The findings underscore the importance of ICH in maintaining cultural continuity and identity in the diaspora, providing insights into the inclusive heritage conservation practices in migrant contexts. This research highlights two key insights: first, the essential role that ICH, along with the associated crafts and movable objects, plays in constructing new homes in the diaspora, particularly in how these items serve as the carriers of cultural identity and continuity; and second, the symbolic significance of Syrian homes, especially their interior designs and decorations, as reflections of a blend of sociocultural practices that Syrians are committed to preserving.

Keywords: intangible cultural heritage (ICH); home-making practices; Syrians in diaspora; migration studies; Liverpool; UK

Citation: Alsalloum, A. Building Home in Exile: The Role of Intangible Cultural Heritage, Crafts, and Material Culture Among Resettled Syrians in Liverpool, UK. *Architecture* **2024**, *4*, 1020–1046. https://doi.org/10.3390/architecture4040054

Academic Editor: Johnathan Djabarouti

Received: 8 July 2024
Revised: 21 October 2024
Accepted: 8 November 2024
Published: 12 November 2024

Copyright: © 2024 by the author. Licensee MDPI, Basel, Switzerland. This article is an open access article distributed under the terms and conditions of the Creative Commons Attribution (CC BY) license (https://creativecommons.org/licenses/by/4.0/).

1. Introduction

Formal and ordered appearance, ritualized arrival and entry, gardens, spacious terraces, and the resulting unneighborly interactivity form a series of tell-tale details and illustrate how architecture gives visibility to the cultural classification of architectural history [1].

For the past decade, Syrians have been striving to find stability either within their country or abroad. By 2021, approximately 28,000 Syrians had relocated to the UK, establishing new homes [2]. The UK government introduced the Syrian Vulnerable Persons Resettlement Scheme in 2014 to offer protection to the most vulnerable Syrians. The scheme concluded in 2021, with the majority of Syrians resettled through government assistance and a smaller number through community sponsorship [3]. Those resettled under this scheme arrived with UNHCR recognition as refugees and were provided with pathways to permanent residence and citizenship. They also benefited from secure travel arrangements, initial reception and support, housing, English language classes, employment assistance, healthcare, and other essential services [4].

Despite the loss or risk of destruction to their built heritage, which holds ancestral stories and shapes their lives, Syrians have successfully preserved and carried with them their intangible cultural heritage (ICH). This encompasses their inherited knowledge and daily practices that define their identity. Although "making home is arguably one of the most universal and, at the same time, social, cultural, and place-specific processes that characterize human life" [5], the impact of ICH on displaced people and their efforts to create new homes and adapt to new settings has not been thoroughly investigated. Specifically, how ICH traditions and practices influence the tangible aspects of interior design and decoration, as well as the style, materials, and objects used within the home for both decorative purposes and daily living has not been addressed. To provide an overview of the ICH practices in Syria, the following list presents examples of ICH elements from the country. These elements are categorized according to the UNESCO-defined domains [6] and are documented by the Syria Trust for Development [7] (Figure 1).

Figure 1. Intangible cultural heritage traditions practiced in Syria. Top left: traditional Laurel soap; top right: traditional copper carving. Bottom left: traditional lithography; bottom right: Arabic coffee making. All these traditions are practiced in Syria. Courtesy of the Syria Trust for Development 2021.

1. Oral traditions and expressions: Examples include mūlawīyah (traditional folk-lyric poetry) and Syriac Christian music.
2. Performing arts, such as Circassian dances [8] and shadow-play theater.
3. Social practices, rituals, and festive events, including the celebration of Eid al-Fiṭr and the rituals associated with traditional coffee-making and drinking [9].
4. Knowledge and practices concerning nature and the universe: An example is the role of an al-ʿaṭār, a practitioner who dispenses traditional herbal remedies and is commonly found in specialized shops within historic marketplaces.
5. Traditional craftsmanship: Examples include rug-making and weaving, as well as ʿajamī painting [10].

The notion of home is a universally relatable concept, often associated with comfort and safety. Home has been defined in various ways, such as a socio-spatial unit, a psycho-spatial condition, and a "warehouse" of emotions and sentimental attachments [11]. The concept of home differs significantly from that of a house, and similarly, the meanings of space and place vary. According to Smyth and Croft (2006), space becomes meaningful "with the life that occupies it", changing both materially and spiritually with the presence of aggregated value in that location [12]. Perkins et al. (2002) distinguish between a house and a home by stating that while space is something we live in, a home is something that is conceived or conceptualized. The homeowner arranges their home based on needs and personal taste, and the inhabitants adapt their homes through decoration and personalization [13]. This allows residents to express their personalities, making the interior and contents of their homes a reflection of themselves [14].

There is a growing body of scholarship that systematically examines how displaced individuals create, reproduce, and re-enact home in various circumstances [15–19]. Scholars often highlight the dual meaning of home as both "a bounded place" and "a meaningful and emotionalized kind of relationship with place" [20]. International organizations, particularly the United Nations Refugee Agency (UNHCR), have initiated several campaigns to understand what home means to refugees and how they temporarily resettle in camps, which sometimes become long-term residences. For instance, the #WhatHomeMeans campaign captures images from refugees to explore "what does home mean when you've been forced to flee yours because of war and persecution; what makes somewhere new a home?" [21]. Existing research has primarily examined how displaced Syrians engage in home-making practices during their temporary stays in neighboring countries, such as Jordan, Egypt, Lebanon, Iraq, and Turkey [22,23]. Additionally, the larger body of studies has explored how Syrians engage in home-making practices during their prolonged and fragmented journeys out of Syria, which involve experiences of home loss. These journeys entail creating a sense of home in challenging displacement conditions and include participating in new social, cultural, or economic activities, recalling the memories of home, and engaging in struggles over rights and belonging. These home-making practices continue even in the absence of family and community connections and the perception of Syrians as homeless [24].

Azraq and Zaatari camps have been the focus of several studies investigating aspects of home-making and place-making among Syrian refugees. One study by Albadra, Coley, and Hart (2018) assessed the indoor environmental quality of shelters and found that the unsatisfactory indoor conditions led many households to use scrap materials to expand their living spaces, creating outdoor areas like courtyards [25]. This adaptation not only improved their living conditions but also demonstrated their resilience and resourcefulness, as these expansions utilized traditional building knowledge passed down through generations. Similarly, Kirk et al. (2018) focused on life inside standardized modular shelters in the Zaatari camp, revealing that interior decorations served as more than just functional elements. They acted as psychological escapes and coping mechanisms for the residents, who often upcycled old materials to create décor that helped maintain a sense of dignity and pride amidst scarcity [26]. This practice highlighted the importance of preserving cultural identity and personal agency in displacement settings. Another study by Paszkiewicz and Fosas (2019) examined how different levels of refugee agency affected mental health and overall wellbeing in both the Zaatari and Azraq camps. Their findings underscored the importance of providing refugees with greater control over their living conditions, which humanized the refugee experience and enhanced their mental health. This study suggested that resettlement organizations should focus on improving refugee homes to meet the unique needs and experiences of the displaced, rather than merely providing basic shelter [27].

Dalal's (2017) research on interior decorations and place-making in the Zaatari camp complemented the above findings as it highlighted the diverse decorative techniques and styles employed by Syrian refugees from different regions. Despite originating from the same country, these refugees brought unique cultural elements to their shelter decorations, reflecting a rich diversity within the camps [28]. Dalal's study provided valuable insights but also indicated the need for more detailed comparisons and standardized categories to fully understand the impact of such practices. Zibar et al. (2022) examined the concept of home among Syrian refugees in the Kurdistan Region of Iraq, revealing that home is composed of intertwined tangible and intangible bonds. The tangible aspects include the physical spaces and the people rooted in various temporal and geographical contexts. Meanwhile, the intangible bonds involve the emotional connections that link one's former self with the positive emotional states associated with their previous home [29]. These studies collectively emphasize the critical role of environmental and personal agency factors in the well-being of displaced individuals. They highlight the creative and resilient ways refugees adapt to challenging circumstances, using inherited traditional knowledge and limited resources to recreate a sense of home and preserve their cultural identity in exile. This is particularly evident in how they transform their living environments through decoration, crafts, and small objects. Moreover, they utilize this traditional knowledge to shape and extend spaces, creating both movable and immovable elements that reflect their socio-cultural identity.

Another body of research on the notion of belonging and the creation of home was conducted in distant places such as Europe, Australia, and North America [30–32]. However, these studies have predominantly focused on migration rather than ICH and the associated crafts. For instance, Shamman et al. (2022) in their book "Migration, Culture, and Identity" explore the dynamic processes of making, remaking, losing, and reviving homes through migration, questioning what it means to create a home away from home. They ask how spaces of resettlement reflect both the continuation of old homes and the emergence of new experiences [33]. Although some research has touched on the cultural aspects, there remains a scant understanding of the comprehensive impact of ICH traditions. For example, the study by Hashem et al. (2022) highlights the importance of recording life stories, cultural and political memories, and archiving oral histories to document and establish a counter-narrative led by the displaced. When examining the meanings and experiences of home for displaced individuals, Hashem et al. (2022) argue that making home for irregular migrants from the Global South is often difficult and sometimes unattainable within the current climate in Britain, mainly because of the perception of the particular sociocultural groups [34]. Creative expressions, such as visual arts and music, can provide meaningful ways to experience home and home-making [35]. These practices transcend territorial boundaries, addressing and bridging emotions across cultures, languages, and generations. For instance, the OneLoveKitchen project in Athens exemplifies how migrants and refugees can enrich local culinary culture with their recipes and practices of sharing, care, and solidarity, thereby constituting a dynamic and transformative kind of ICH [36].

While literature on how refugees create homes exists, the focus on ICH has been limited. This is crucial, as ICH maintains a sense of identity and facilitates understanding and communication between different cultural groups. Furthermore, significant research on ICH related to refugees often pertains to camp settings. For example, international laws and their relationship with ICH have been reviewed to highlight the importance of safeguarding refugee heritage during the journey from persecution to resettlement [37]. Efforts by organizations like Homeland Document to archive war-related testimonies and traditional crafts contribute to reshaping individual and collective memories. Studies show that such documentation is vital for understanding and preserving the ICH of displaced communities [38]. Nibal (2024) argues that Syrian ICH embodies a unique and intrinsic link between physical and non-physical heritage in their homeland. This connection is evident in cultural spaces, like mosques, churches, bazaars, and khans (caravanserais), which not only represent official heritage through their monumental structures but also serve as public venues for traditions and crafts. Furthermore, Syrian ICH carries a significant emotional dimension, making it particularly sensitive and vulnerable to the pressures of modernity and economic growth. This emotional connection underscores the importance of preserving such heritage against the encroaching forces of development and commercialization [39]. Accordingly, the study of how migrants construct their domestic living environments and the challenges they face in creating a sense of home is increasingly gaining prominence in cross-disciplinary research. Housing issues provide valuable insights into their transnational social engagement and the evolving boundaries of identity and belonging. However, there remains a significant gap in the research on the role of ICH in the home-making practices of resettled Syrians in the UK. Specifically, there is limited understanding of how Syrians have applied their inherited knowledge and traditions to redecorate their homes and use crafts and objects in everyday life to recreate a living environment where they feel at home again. This gap is particularly important in light of the UK's recent ratification of the 2003 UNESCO Convention on Safeguarding ICH, following a decade-long delay.

The situation of displaced Syrians in the UK offers a unique case for exploration. Liverpool, with its rich history of immigration and diverse population, has become a significant destination for resettled Syrians [40,41]. Despite Liverpool's vibrant cultural landscape, there is limited academic focus on the interactions within these groups [42]. Existing studies on migrant ICH in the UK and the integration of Syrians specifically often overlook the everyday practices and cultural elements that shape their lives' particular domestic spaces. My unique perspective as a cultural heritage researcher, an architect, and a British Syrian living in the UK offers a distinct vantage point to explore how Syrians use elements of their ICH to recreate homes and adapt to their new environment in Liverpool. The primary objective of this research, therefore, is to raise awareness of Syrian heritage within the diaspora by focusing on both tangible and intangible cultural heritage elements within the domestic spaces. The research is designed around the following main questions:

1. How have resettled Syrians in Liverpool, UK, adapted their domestic environments to reflect their intangible cultural heritage?
2. What are the defining characteristics of a "Syrian home" in the context of displacement, and how does this home represent and support the practices of living heritage?
3. Which traditional artifacts, cultural customs, practices, and everyday objects—such as crafts and movable items—hold the most sentimental value for Syrian migrants, and how do these elements help them overcome the challenges of establishing new homes?

This paper responds to this journal's special issue's focus on the dynamic relationship between intangible cultural heritage (ICH) and built heritage conservation by examining how resettled Syrian communities in Liverpool integrate traditional crafts and cultural practices into their new domestic environments. By exploring how these practices contribute to both the safeguarding of cultural identity and the adaptation of built spaces, this study offers insights into inclusive conservation and the transmission of intangible heritage within the process of creating new homes in exile. Thus, this research contributes to the broader interdisciplinary discourse on migration, intangible cultural heritage (ICH), home, place-making, and integration. It seeks to challenge the prevailing narratives and promote a deeper understanding of the complex experiences of domestic spaces within migrant communities in the UK, specifically Syrians. This is accomplished through a hands-on exploration of the homes of Syrians settled in Liverpool, UK. Conducted in their native Arabic, the interviews took place in their authentic home environments, supported by a thorough literature review and grounded in an original, significant topic. This research makes a valuable contribution to the ethnographic studies of home-making practices in the UK, with a methodical approach that is clearly detailed and presented in an accessible manner. High-quality images further illuminate key aspects of the study. Overall, this study addresses an underexplored area of ethnography and place-making and enriches the existing body of knowledge. The article focuses on ICH by examining how traditional objects are used to facilitate key cultural practices, such as religious rituals, celebrations, and hospitality customs. While references to tangible items—such as traditional objects displayed in cabinets, Arabic calligraphy in living rooms, and tables adorned with traditional sweets—are insightful, the primary aim is to show how these objects support and sustain ICH practices. These items serve as carriers of intangible traditions, embodying deeper cultural meanings and reinforcing the rituals, customs, and social interactions that define the ICH of Syrians in exile.

2. Materials and Methods

As a female British Syrian researcher and architect based in Liverpool, I had the unique opportunity to visit the homes of nine families and conduct interviews with 22 individuals. These interviews were conducted in the participants' native Arabic language, allowing for a deeper connection and more authentic responses. With their consent and alongside an experienced filmmaker, we documented six of these homes—four houses and two flats—through photographs. This study employs a methodology combining semi-structured in-person interviews, observations, and photographs to collect data directly from the interviewees' homes. Ethical approval was obtained from the University of Liverpool's Research Ethics Committee (reference: 11100). Interviewees were mainly recruited through the Syrian British Cultural Centre (SBCC) in Liverpool. Eligible interviewees were those aged 18 or older, who had resided in the UK for at least four years as of 2022, along with their families. The geographical focus was Liverpool due to its significant Syrian community and the accessibility for the researcher and filmmaker. The University of Liverpool Higher Education Innovation Fund (HEIF) funded the research, with a limited timeframe for completion.

2.1. Interviews

The semi-structured interviews were divided into three sections:

1. General participant information: Collecting demographic details such as age, gender, educational level, and the duration of UK residency.
2. Awareness and adaptation of Syrian heritage: Exploring interviewees' awareness of the Syrian heritage, sentimental items they brought with them, and how they adapted their UK homes to create a sense of belonging.
3. Traditional practices: Focusing on the traditions that interviewees maintain within their homes and the adaptations they have made to continue these traditions in the UK, whether through practicing their customs or utilizing specific crafts and movable items similar to those they used in Syria.

Interviewees provided written consent, with nine agreeing to be identified by their first names, while others chose anonymity. To respect the privacy of all interviewees, the names used in this paper are pseudonyms, as preferred by the majority of the interviewees. Information sheets detailing the project, anonymity assurances, and data usage protocols were given to all interviewees. Debriefing letters, including useful contacts for assistance, were also provided after each interview. The interviews were audio and video recorded, with durations ranging from 30 to 40 min. Emotional responses during the interviews were noted, especially when the interviewees recalled loved ones lost or family members still in Syria. Conducting the interviews in Arabic and being Syrian myself facilitated a deeper connection and understanding. Follow-up contacts were made to check on the interviewees' well-being post-interview. Considering the open-ended nature of the questions and the diverse issues raised by interviewees, an analogical research approach was employed. This approach uses analogies to gain insights into one field by drawing parallels from another, a method well-regarded in various disciplines, including cognitive science and architectural design [43]. The analogical process was guided by three primary coerces: similarity, purpose, and structure [44]. When interviewees were asked about traditional practices, the term "Intangible Cultural Heritage (ICH)" was explained with examples, as it was unfamiliar to some. Building on this framework and following methodologies outlined by Kvale [45] and Brinkmann and Kvale [46], the interview data analysis proceeded systematically as follows:

1. Recording and transcription: Interviews were conducted in Arabic, recorded, and thoroughly transcribed. The transcriptions were then translated into English.
2. Themes identification: Key points were annotated during transcription to identify themes. The data were subsequently organized in alignment with the study's main research questions and the interview structure.

3. Thematic analysis: Utilizing methodologies outlined by White et al. [47], recurring themes were identified, focusing particularly on the interviewees' perceptions of Syria's heritage and the practice of ICH in their houses.
4. Coding: An in-depth line-by-line coding method was applied [48], shaped by the core research questions and themes previously identified.
5. Narrative analysis: Central narrative components were highlighted and aligned with the principal findings of this study. This approach provided a comprehensive understanding of ICH and its associated themes [49].

The integration of the analogical research strategy with narrative analysis enhanced the reliability, validity, and applicability of the study. This comprehensive approach ensured the authenticity of the interview insights while addressing the inevitable limitations related to the time and availability of interviewees. Consequently, this method facilitated a deep understanding of the interviewees' experiences and perspectives, highlighting the strength and intricacy of the research. This qualitative study involved in-depth interviews, photographs, and recordings of social and cultural events, all conducted in the respondents' authentic settings. By focusing on their homes and regular social gathering spots, the study aims to gain deeper and more genuine insights into the cross-disciplinary research on ICH, ethnography, and home-making in exile.

2.2. Observational Study

Observational methods in qualitative research are pivotal for gaining deep insights into the lived experiences of individuals within their normal settings. These methods allow researchers to capture the nuances of everyday life and the subtle ways in which cultural practices are embedded in daily routines [50,51]. According to Spradley (1980), participant observation is essential for understanding the social and cultural contexts of the studied group, offering a comprehensive view of their interactions with their environment [52]. In the context of migration studies, observational methods are particularly valuable for exploring how migrants adapt their new living spaces to reflect their cultural heritage. This approach has been effectively employed in various studies to understand the home-making practices of displaced populations [53,54]. For example, Boccagni (2017) used observational techniques to explore how migrants personalize their living spaces [55], revealing the importance of material culture in the process of home-making and identity construction. This study focused on living rooms, reception rooms, dining rooms, and kitchens. The observations included interior order, household furniture, decorations, and displays. The aim was to understand how these spaces were adapted using craft objects and household items that are related to ICH practices to create a sense of home. The study highlighted the influence of ICH traditions on home physical settings and the interplay between personal actions, sociocultural traditions, and home-making in the diaspora.

Interior order and household furniture: Observations included noting the arrangement and types of furniture present in these key areas. This helped in understanding the functional adaptations and aesthetic choices influenced by ICH traditions. For instance, traditional seating arrangements, such as floor cushions in living rooms, were observed as a nod to Syrian customs in some regions.

Decorations and displays: This study also focused on various decorative elements, such as wall hangings, artifacts, and other displays. These items included traditional crafts, family photographs, and religious symbols. The presence and placement of these items provided insights into how cultural identity is maintained and expressed within domestic spaces utilizing tangible items.

Sociocultural traditions and home-making: A key aspect of the observation was to understand the interplay between personal actions, sociocultural traditions, and the physical environment of the home. This included observing how spaces were used during social gatherings, family rituals, and daily routines, thereby highlighting the role of ICH traditions in home-making and the maintenance of cultural practices in the diaspora.

2.3. Limitations of the Methodologies

The study's geographical focus on Liverpool and the limited number of interviewees (n = 22) may not represent the broader UK Syrian community. Additionally, the research primarily focused on craft objects and household items directly associated with ICH practices rather than conducting a detailed investigation of private and public spaces within the house. Observational studies were limited to common areas such as living rooms, reception rooms, dining rooms, and kitchens, which may not fully capture the extent to which ICH traditions have influenced the integration of craft and household objects throughout the entire home. Moreover, while observations provided valuable insights, they were constrained by the subjective interpretation of the researcher, and the presence of the researcher might have influenced the behavior and environment being studied.

3. Results

This study's findings reveal that after resettling in the UK, Syrians have actively maintained and practiced their ICH, particularly traditions tied to daily life and family practices from their homeland. The continuation of these traditions has been deeply connected to the preservation of both tangible heritage—in the form of household items and crafted objects—and cultural memory, as these objects evoke and sustain connections to their Syrian identity. By purchasing and incorporating traditional items, such as those used for preparing food or decorating their homes, Syrians in the UK recreate a sense of belonging that ties them back to their motherland. These items, often handmade and passed down through generations, serve not only as physical objects but also as vessels of memory, reflecting the living heritage of a community navigating displacement.

The integration of these traditions into new living spaces has required adaptations, reflecting a negotiation between cultural preservation and the practicalities of life in the UK. Interviewees highlighted positive aspects of UK society that helped them adjust while also addressing the challenges they faced in safeguarding their heritage. Practices centered on home decoration and family rituals were seen as crucial for maintaining cultural resilience and identity, helping individuals adapt to life in the diaspora by transforming their new homes into familiar, meaningful spaces. Interviewees emphasized the importance of safeguarding these practices, both for themselves and future generations. The continuation of traditional activities within the home—whether through cooking, hosting social gatherings, or practicing religious rituals—emerged as essential to their emotional well-being and sense of belonging. By doing so, they not only sustain their identity but also contribute to a broader dialogue on the role of heritage in fostering resilience and adaptability. The adaptation of domestic spaces demonstrates how cultural identity is expressed through the built environment, as rooms and decorations are modified to preserve and represent cultural traditions. These adaptations are a form of architectural continuity, illustrating how migrant communities negotiate new environments while maintaining strong links to their heritage.

The results of this study are organized into five sections: First, an outline of the interviewees' demographic profiles. Next, the study examines the items they brought with them, if any, when they first arrived in the UK. This is followed by three thematic sections: traditional practices, challenges, and adaptations; representing home and preserving cultural continuity; and the importance of safeguarding Syrian heritage.

3.1. Demographic Outlines

As previously stated, this study involved interviews with twenty-two individuals conducted between March and May 2022. All interviews were conducted face-to-face. Among the interviewees, sixteen were women and six were men (Figure 2). The interviewees were aged between twenty-seven and seventy years old. Each participant had finished at least secondary education, with seventeen having completed higher education. Their current professions in the UK include housewife, specialist doctor, school teacher, restaurateur, businessperson, and student. All interviewees had lived in Syria during and prior to the war and relocated to nearby host countries, including Jordan, Lebanon, and Turkey, between 2012 and 2017. Most of the interviewees had been residing in the UK for a period ranging from four to ten years. They primarily came from three main regions in Syria: Damascus, Horan, and Aleppo. This diversity of backgrounds allowed for a nuanced exploration of how Syrian ICH manifests in their new homes.

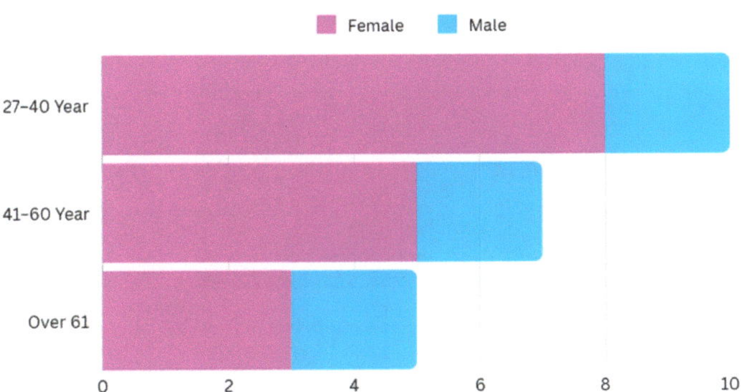

Figure 2. Interviewees' gender and age group.

3.2. Cherished Possessions Brought from Syria

This study found that the majority of Syrian males initially migrated alone, and after they settled, they arranged for their families to join them. Some men traveled by sea, while others trekked overland through Europe. Despite the arduous journey, some managed to bring small items that reminded them of home. Most of these items were small and not necessarily valuable in economic terms, but they held immense sentimental value. They reminded the individuals of dear family members, close friends, or significant events and special occasions in their lives. However, some were unable to bring even small items, as they had to travel through several countries before resettling in the UK, leading to the loss of these cherished possessions. For these individuals, only memories remain. The following are some examples of the items they brought.

For instance, Momen could not bring anything with him as he lost most of his belongings during the trip to the UK. When his family joined him after four years, he requested an old Quran, a gift from a dear friend, which is now his only keepsake from Syria. Gazi brought a traditional Islamic prayer bead string, culturally and socially significant, often held by men during social gatherings. These beads, handcrafted locally in various regions of Syria, vary in shape and size. They are used during social gatherings, traditional dances like Debkeh, and prayers. He also brought a special ring worn by his father and grandfather, signifying marriage. Similarly, Sara brought two traditional wedding gifts from her parents: a small bottle of perfume and a special candle. The perfume, a gift from her late father on her wedding day 18 years ago, evokes memories of him every time she sprays it. She said, "Whenever I spray a very little bit from this perfume, I remember my dad. RIP". The candle, a gift from her mother, remains unused as a cherished memento of her family still in Syria. Fatema brought traditional clothes, including an Abaya, which is part of her cultural attire. She mentioned that her husband, who arrived before them, advised them to bring traditional coffee and tea pots, items that were not available in the UK market at the time. Kholod brought photo albums, selecting a few photos from each stage of her life, as well as her husband's and children's lives. She said, "When I came, I was scared and only had my handbag, but I brought a photo from each stage of our life. This is our life in one album", reflecting on the significance of these memories (Figure 3).

Figure 3. Cherished possessions brought from Syria.

It is clear that the items brought by the interviewed Syrians are deeply connected to their ICH traditions. These items, whether religious (Quran), artisanal (traditional bead strings), or socio-cultural (traditional coffee and tea pots, clothes), represent and facilitate the continuity of their cultural practices in their new environment. Table 1 illustrates the variety of items carried by the Syrian migrants to the UK and their emotional or cultural importance.

Table 1. Cherished personal items brought by interviewees from Syria and their symbolic significance.

Item Type	Symbolic Value
Religious (Quran)	A reminder of dear friends in Syria.
Cultural (prayer beads, ring, special coffee and teapots)	Connections to communal traditions and hospitality.
Family heirlooms (perfume and candle)	Evoking memories of family back in Syria and/or lost family members.
Family photographs	Photos representing different life stages, "our life in one album", as the interviewee said.
Traditional clothing	Represent Syrian identity.

3.3. Traditional Practices, Challenges and Adaptations

One of the main challenges for many interviewees was the difficulty of acquiring household items similar to those in Syria upon their initial arrival, particularly about nine years ago. One participant said, "We did not have anything to remind us of home". Some interviewees highlighted the challenges and the subsequent improvements in availability, noting that initially they had very little. However, as the Syrian population in the UK grew, so did the availability of traditional items. Samer said, "Special items for traditional celebrations, such as the Erk Souse drink, which we consume during Ramadan, were not available before, but now we have almost everything as the market offers it all. In 2016, I had to go to London to buy an Arabic calligraphy tapestry. But now, in Liverpool, there are traders who sell these items and household goods".

Mariam also reflected on the initial challenges and subsequent improvements in availability: "Initially, it was hard to find household items. Now, there are many positives, as food items and traditional household goods are available. For example, we can now find Makdous, pepper paste, and special Syrian pickles, which are different from those available in the UK market. Specialty shops now offer these items, and we have the raw materials to make them at home, which we previously lacked". Over time, the demand for these items increased, prompting some Syrians to start businesses importing Syrian or similar products (food and household items) from Lebanon, Turkey, or Jordan to sell in specialty shops. These items have also become popular among other nationalities.

Some participants noted that celebrating Eid al-Fitr and Ramadan was initially difficult because non-Syrians were not aware of these traditions, and the interviewees had to explain why they were fasting. However, there has been much greater awareness now, and people have accepted living in a multicultural environment. One of the key challenges interviewees noted was their initial lack of knowledge about the norms and culture in the UK. Zein stated, "We are still learning and integrating. Now we have established a good community and a band to sing traditional songs on special occasions. Some UK communities now attend and learn about our traditions". Another participant said, "We have made some changes to our traditions. In Syria, we stay up late during Eid and other special occasions, but here in the UK, we respect that by 11:00 p.m., we need to be quiet for the neighbours".

Another challenge highlighted by several interviewees was the size of living spaces in the UK, which was notably different from their previous living environments. As a result, many Syrians have become creative in decorating their new homes. Fatema said, "We are unable to change the house and its size, which is why we change its decoration and interior". In Syria, it is common to have balconies where people sit and enjoy their coffee with some plants around. While houses with balconies are not common in the UK, Syrians have adapted by focusing on interior decoration to create a familiar environment. Living in a house similar to their homes in Syria was extremely important for interviewees, with space being a primary concern. One participant said, "We like to pray with our children and the whole family together, but sometimes we have to move furniture to fit everyone in one room (five children and two parents)". Another male interviewee shared, "In Syria, we used to have larger homes and typically received male and female guests separately. We preferred for males to sit together to discuss social and business matters, while females would gather separately. In the UK, this was challenging due to the smaller house sizes, but we got approval to build a conservatory, which helped solve this issue".

Another common house type in Syria is the courtyard house, typically found in older cities. Residents would sit in the courtyard, which often featured a small garden and a water fountain. Additionally, there were separate rooms designated for receiving male and female guests. Khoulod shared, "I discovered a house model resembling the traditional courtyard houses, with a courtyard featuring a fountain and greenery, which served as both a sentimental and climate equalizer. Unfortunately, most of these houses have been destroyed during the war, particularly in Old Aleppo. I was delighted to find this model because it mirrors the old courtyard houses in Old Damascus" (Figure 4).

Figure 4. A model incorporating elements of a traditional Syrian courtyard house, displayed in a Syrian family home in Liverpool, UK.

In some parts of Syria, particularly the southwest region of Horan, people sit on the ground and use special furniture for their living rooms. After migrating to the UK, they maintained this tradition by creating rooms styled with floor seating, continuing their custom of sitting on the ground for meals and social gatherings. Omar said, "In Horan, we sit on the ground on special furniture and eat while sitting on the ground". Moamen, also from Horan, added, "Sitting on the ground is part of our customs and reminds us of happy days in Syria. When we first lived here, we bought sofas, but the room felt smaller. After we switched to floor seating, we could invite more friends for dinner, sometimes hosting around 30 men" (Figure 5).

Figure 5. Traditional furniture in a sitting room for floor seating in one of the Syrian homes in Liverpool, UK.

One female participant from the same region said, "We still eat while sitting on the ground. This is a tradition we've maintained in the UK so our children can pick it up and continue practicing it. Once, we bought a dining table, but it didn't feel right, so we replaced it with our traditional floor seating" (Figure 6).

Figure 6. A style of traditional home seating and traditional coffee setup in one of the Syrian homes in Liverpool, UK.

Despite the challenges posed by smaller homes or flats, interviewees mentioned that houses in the UK are warmer in winter due to their structure and size. Additionally, they felt welcomed by the UK community. One participant noted, "The community here is very welcoming, and they are keen to learn about our culture. They participate in our food, order catering, and even join in dancing". Another participant added, "We appreciate that in the UK, we are free to practice our traditions in a multicultural society, and the UK community even participates with us, trying our food and joining in our dances". Table 2 below outlines the practical challenges of recreating Syrian home environments and the solutions adopted by the interviewees.

Table 2. Challenges in recreating Syrian homes and adopted solutions.

Challenge	Adaptation	Traditions to Be Continued/Reason for Change
Difficulty acquiring household items similar to those in Syria upon initial arrival	Establish businesses to import items	Everyday life practices, decorations, hospitality
Initial lack of awareness among the UK community of Syrian traditions	Continue traditions; explain to colleagues	Social and festival traditions such as Eid
Initial lack of awareness among Syrians about UK traditions and cultural environments	Learn and integrate with non-Syrian communities.	Practicing traditions like fasting during Ramadan and celebrating Eid
Small living spaces compared to homes in Syria	Creative traditional decoration, building conservatories, Majlis (floor seating)	Hospitality, more space for family prayer at home, gender-specific spaces, privacy, sitting on the ground to maintain social and cultural traditions with large gatherings, eating by sitting on the ground as a custom

3.4. Representing Home and Preserving Cultural Continuity

The findings of this study indicate that Syrians are keen on representing their homes in ways that reflect their identity while balancing both aesthetics and functionality. This is crucial for maintaining cultural continuity on both individual and family levels, as it reminds them of their homeland. Additionally, it plays a significant role at the community and social group levels, acting as an expression of identity. For example, Fatima expressed her desire to maintain cultural traditions in her home decor, stating, "I like to decorate my house with Ramadan-inspired decorations and Arabic calligraphy, which remind me of my parents' home where I was raised. This is also important for hospitality and inviting friends". She also brought a small mosaic tray and laurel soap, traditional crafts from Aleppo (Figure 7).

Figure 7. Table with traditional items and sweets during Eid, Liverpool, UK.

This study also found that Syrians are keen on preserving their social life in the UK, which provides them with a sense of home. They have been diligent in acquiring specific items to prepare and serve their traditional food and drink. Khoulod said, "We are trying to live a social life similar to what we had in Syria. For example, we practice subhiyah, when a group of ladies gather to have Syrian coffee together in the morning or sometimes extend it to a joint breakfast. I'm particular about using traditional coffee cups and trays when serving my guests. We strive to continue our social traditions as we did in Syria".

Hospitality remains a key tradition for Syrians, and it continues to be an important part of their lives in the UK. Moamen shared, "I remember my late father inviting family and friends for dinner in Syria, especially for dishes like Mlehy. I want to continue this tradition and hope my children do too. We want our homes in the UK to bring us back to those happy memories" (Figure 8).

Figure 8. Mlehy food served on the ground in a traditional setting in one of the Syrian homes in Liverpool, UK.

Including non-Syrian friends in these hospitality traditions is also significant. Zein said, "My non-Syrian friends enjoy Syrian coffee in traditional cups when they visit. They are curious about the items my mother displays in a cabinet, such as mosaics and Arabic calligraphy, similar to those we had in our house in Syria". Majed's mother added, "Having a special cabinet to display traditional items like mosaic, silverware, and other crafts is a key decoration in Syrian homes. While some display these items without a cabinet, we have maintained this tradition here in the UK" (Figure 9). This study also observed the display of traditional tapestries and intricately engraved Arabic calligraphy as a distinctive aspect of Syrian homes, creating a unique and culturally rich environment (Figure 10). Another notable tradition is the use of custom-made curtains, often ordered from special tailors, which feature three or four layers. These types of curtains were common in Syrian homes and flats, valued for their aesthetic character and privacy (Figure 11).

Another aspect Syrians continue to cherish is the use of frankincense and scents to create a special atmosphere in their homes. Luna, who started a small business crafting items from raw materials sourced from Syria, shared her story: "My late grandmother used to make similar items for us. I started this as a hobby, gifting my friends on special occasions, and they encouraged me to expand. Now, I make candles, soaps, and bath bombs using Damascus jasmine and oud, as well as jewelry with traditional Syrian embroidery" (Figure 12). Nada also started a business importing items from Syria and other countries like Lebanon, Turkey, and Jordan. Her business, which has been growing for two years, meets a demand for traditional Syrian items, such as tapestries, laurel soap, and oud. She stated, "I noticed demand not only from Syrians but also from other nationalities. I even

import Muna items, traditional preserved food from Syria, which people ask for frequently"
(Figure 13).

Figure 9. Traditional items displayed both inside and outside a cabinet in two different Syrian homes in Liverpool, UK, along with a close-up view of some of the items.

Figure 10. Arabic calligraphy items in the living rooms of three different Syrian homes in Liverpool, UK.

Figure 11. Curtains in two different Syrian houses in Liverpool, UK.

Figure 12. Handmade products from raw material from Syria by one of the interviewees, Liverpool, UK.

The Syrian community interviewed in this study has embraced various forms of cultural expression to navigate new situations and discomforts. This includes joint rituals of grieving and mourning, as well as collective practices of bonding and remembrance. Celebrations such as births, marriages, and other festive events are also significant. One participant shared, "We have practiced these traditions since we were children. Standing by each other in times of joy and grief is deeply rooted in Syrian culture, and we are keen on maintaining this in the UK. For example, when someone loses a close family member, we observe three days of mourning, called Ajer, where people offer help and support, ensuring the bereaved person does not feel alone. We also prepare traditional coffee for the mourning period". Omar added, "On Eid, we wake up early, prepare traditional coffee, and make sweets like mamool. Initially, it was hard to find the ingredients, but now they are more accessible. Our community has grown, and we have good friends who join us in wearing traditional clothes, which makes us feel at home" (Figure 14). In another instance, a female participant shared, "When I gave birth to my daughter, one of my friends would come daily to cook and clean for me. This made me feel at home". These small adjustments to their homes and social lives are crucial for maintaining their deeply rooted traditions in exile. Syrians have been keen on acquiring items that connect to their intangible cultural heritage (ICH), such as those used in food preparation, hospitality, and interior design, including curtains, mosaic pieces, and Arabic calligraphy. These items help fill cultural gaps and represent their identity while adapting to a new environment.

Figure 13. Items exhibited in a traditional Syrian shop, Liverpool, UK.

Figure 14. Traditional Syrian sweets in one of the interviewees' houses during Eid, Liverpool, UK.

3.5. The Importance of Safeguarding Syrian Heritage

Maintaining and passing down these traditions is vital for safeguarding Syrian heritage and ensuring that future generations remain connected to their roots. All interviewees stressed the importance of continuing specific traditions, particularly those related to home decor, hospitality, and cultural practices. To ensure their children stay connected to their heritage and learn Arabic, the community has established two Arabic weekend schools, which have also attracted students from other Arabic-speaking nationalities. One participant explained, "One of our traditions is for children to speak with their grandparents, aunts, and uncles, especially during Eid and other special occasions. Thus, it's important that they learn Arabic, which is why we send them to weekend Arabic schools". Another participant shared her efforts to create a family environment by watching Arabic documentaries with her children. Additionally, another participant stated that he sings traditional songs with his three children using a simple instrument called a daf, wanting his children to remember these songs and be reminded of home (Figure 15).

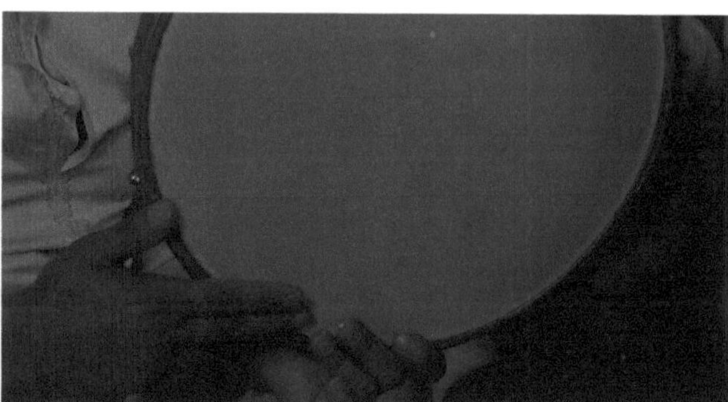

Figure 15. One of the interviewees playing the daf and singing with his children, Liverpool, UK.

Fatima described how her family preserves cultural traditions: "Every Friday, we play Quranic recitations, which create a special environment. My children enjoy it because it reminds them of home. My youngest, who is seven, helps me decorate the house and make traditional Syrian cookies". Majed, a young man from Damascus, shared his experience of losing all his belongings during his journey to the UK: "I used to play the oud, and I felt a gap in my life without it. I bought a new oud, but it was uncommon here in the UK, and I couldn't find a teacher. I eventually learned the guitar. Music helped me fill the cultural gap I felt and motivated me to adapt to a new culture while holding onto parts of my own". Table 3 below outlines tangible and intangible cultural heritage items and practices preserved by Syrians in Liverpool, according to this study.

Table 3. Tangible and intangible cultural heritage items and practices among Syrians in Liverpool.

Theme	Tangible Heritage (Physical Objects)	Intangible Heritage (Practices/Meanings)	Cultural Significance/Practices
Religious and Family Heritage	Quran, prayer beads (misbaha), ceremonial rings, heirlooms (perfumes, candles)	Quranic recitations, Ajer (grieving practices), prayers, storytelling	Preserving religious identity, honoring family, and emotional ties to faith and heritage
Traditional Craftsmanship	Coffee pots, tea pots, Abaya (traditional clothing), laurel soap, mosaic trays	Craft-making, use of Syrian soap, handmade decorations	Continuity of craftsmanship, cultural pride, identity expression
Social Practices and Hospitality	Coffee cups, serving trays, floor seating (Majlis), dining sets	Hosting guests, subhiyah (morning coffee gatherings), communal dining	Strengthening community bonds, social life, and hospitality traditions
Festive and Religious Celebrations	Traditional sweets (mamool), Arabic calligraphy tapestries	Celebrating Eid al-Fitr, Ramadan fasting, religious and communal gatherings	Intercultural exchange, maintaining traditions, bonding through food
Home Decoration and Identity	Tapestries, Arabic calligraphy, layered curtains, decorative cabinets for heirlooms	Displaying heritage items and decorating homes with cultural symbols	Expressing cultural identity, connecting to home, aesthetic and functional design
Culinary Traditions	Makdous, pepper paste, Syrian pickles, special cooking tools	Preparing traditional Syrian meals, family cooking rituals	Culinary heritage as a cultural link, food as memory and tradition
Music and Performing Arts	Oud, daf (musical instrument)	Singing traditional songs, playing music with family	Emotional resilience, continuity of performing arts, bonding

Table 3. *Cont.*

Theme	Tangible Heritage (Physical Objects)	Intangible Heritage (Practices/Meanings)	Cultural Significance/Practices
Adaptation to the UK Environment	Conservatories, floor seating, home extensions	Adjusting homes while preserving cultural practices	Balancing UK living conditions with Syrian traditions, resilience in home-making
Transmission of Cultural Heritage	Arabic school materials, traditional children's toys	Arabic language education, teaching religious and cultural traditions	Safeguarding cultural heritage and passing on traditions to younger generations

3.6. Traditions and Associated Items Captured

The results of this study identified various traditional items, movable objects, and crafts used inside Syrian homes in Liverpool to support their family, social, and religious practices. These items reflect several ICH traditions that the Syrian community has maintained during their resettlement in Liverpool, UK. They serve as expressions of identity and help transform their houses into true homes. Based on the interviews conducted, the following ICH elements were observed:

1. Oral traditions and expressions

   - Storytelling and oral history: Shared during family gatherings and community events. This includes transmitting family histories, cultural stories, and personal narratives.
   - Communication of traditions: Explaining the meaning behind cultural items and practices to non-Syrian friends and guests.

2. Performing arts

   - Playing traditional musical instruments, such as the oud and the daf. Some participants highlighted playing music for family and gatherings, maintaining these traditions in the UK.
   - Singing traditional songs: Interviewees reported singing Syrian songs during family gatherings, celebrations, and with their children to pass on their cultural heritage.

3. Social practices, rituals, and festive events

   - Celebrating Eid al-Fitr and Ramadan: These celebrations include fasting, special prayers, and breaking fast with traditional food, all practiced with family and friends.
   - Hosting subhiyah: Morning gatherings where women serve traditional Syrian coffee, creating a social bond within the community.
   - Friday Quranic recitations: Played in many homes to maintain religious practices and traditions, helping create a spiritual atmosphere.
   - Grief rituals (Ajer): The observation of three days of grief where the community offers support, with traditional Syrian coffee served as part of the ritual.
   - Support provided after childbirth, including communal caregiving and baby showers.
   - Maintaining hospitality traditions: Hospitality is an essential part of Syrian social practice, with participants inviting friends and family, including non-Syrians, to their homes, offering traditional food and drink.

4. Knowledge and practices concerning nature and the universe

   - Using frankincense: Syrian homes in Liverpool use frankincense and other scents like oud to create a familiar and comforting environment. This practice is related to their understanding of nature, spirituality, and atmosphere.

- Incorporating natural materials: The use of materials like Damascus jasmine and oud in crafts and home products (e.g., candles, bath bombs) signifies a connection to natural resources traditionally available in Syria.

5. Traditional craftsmanship

- Making traditional crafts: Participants reported making crafts such as candles, soaps, and bath bombs using raw materials from Syria, continuing the tradition of handmade products.
- Displaying crafted items: Many homes featured cabinets displaying traditional mosaics, silverware, Arabic calligraphy, and other handcrafted items, maintaining a strong link to Syrian cultural aesthetics.
- Decorating with traditional tapestries and curtains: Homes are adorned with traditional Syrian tapestries, layered curtains, and other crafts, contributing to the recreation of Syrian living spaces in the UK.
- Making Muna, which involves various techniques for preserving food, such as preparing pickles, jams, and other similar items.
- Use of traditional serving items: Special Syrian coffee makers, cups, and trays are used to serve guests, highlighting craftsmanship linked to both joy and grief rituals.

4. Discussion

4.1. Summary and Interpretation of the Findings

This study has demonstrated the vital role that intangible cultural heritage (ICH) plays in the adaptation of Syrian refugees to life in Liverpool, UK, by examining their home-making practices and the associated use of tangible items and cultural traditions. The findings reveal that Syrians have actively engaged in recreating familiar domestic environments, transforming their new homes into spaces that reflect their cultural identity and allow for the continuation of important social, religious, and familial traditions. The maintenance of these traditions is not only a matter of preserving ICH but also serves as a means of emotional and psychological well-being. Participants emphasized that incorporating traditional objects and practices into their homes helps sustain a sense of belonging and continuity. Items such as Arabic calligraphy, traditional coffee pots, tapestries, and handmade crafts act as vessels of cultural memory and pride, connecting them to their homeland.

This study responds to the key themes in this journal's special issue by highlighting the interaction between intangible heritage and built spaces. Through traditional crafts, interior decoration, and the adaptation of domestic spaces, Syrian refugees in Liverpool continue to express their cultural heritage in a new environment. This contributes to broader discussions on how cultural heritage influences the shaping of sustainable and resilient design practices in the context of migration and displacement.

4.2. Adapting to New Living Conditions

One of the primary challenges highlighted by the interviewees was the smaller living spaces in the UK compared to Syria. As a result, they had to creatively adapt their homes to continue their cultural traditions. These adaptations included rearranging furniture, using traditional floor seating (majlis), and, in some cases, extending their homes with conservatories to accommodate guests and religious gatherings.

Decorative elements such as Arabic calligraphy, laurel soap, Damascus jasmine, and handmade mosaics were integral to creating a familiar environment reminiscent of Syria. These tangible items not only served aesthetic purposes but also carried deep cultural significance, symbolizing hospitality, religious devotion, and family unity. Through these tangible markers of identity, the participants created spaces that were functional while retaining the essence of their cultural heritage.

4.3. Social Practices, Rituals, and Continuity

These findings show that social practices and rituals are essential components of the home-making process for Syrians in Liverpool. Key traditions such as hospitality, religious rituals, and community gatherings were maintained through the use of traditional crafts and objects. For instance, the preparation of traditional Syrian coffee using handmade cups and serving trays was identified as a central ritual during both joyous and sorrowful occasions.

Participants also spoke of the significance of religious practices, such as Quranic recitations and fasting during Ramadan, in maintaining cultural continuity. These practices not only reinforce their connection to their Syrian heritage but also help build community bonds in the UK by including non-Syrians in their celebrations and hospitality practices. The continued use of traditional objects such as prayer beads, incense burners, and ceremonial garments serves as both a physical and symbolic connection to their cultural roots.

4.4. Comparative Analysis and Reflecting on Literature

The outcomes from this study contribute to the broader literature on intangible cultural heritage (ICH) practices, migration, home-making, and the use of objects and crafts in heritage preservation within the UK. While some findings align with Ilcan and Squire's (2022) study [56], which focused on displaced Syrians resettled in the West Midlands, UK, and London, Canada, the interviewees in this research faced fewer challenges. Although both groups described their initial arrival in the UK as extremely difficult, which hindered the creation of a new sense of home, participants in this study emphasized the role of ICH practices and household objects in facilitating their adaptation and integration into UK society.

Previous studies have emphasized the crucial role of cultural practices in maintaining migrants' cultural identities and providing emotional support during resettlement. Baser (2011) [57] and White (2015) [58] discuss how migrants use these practices to create a sense of belonging in new environments. Similarly, Aragonés et al. (2010), Hapsariniaty et al. (2019), and Momade (2022) [59,60] explored how culture influences domestic spaces, showing that dwellings become homes through an active modification of the surroundings. However, discussions on the specific ICH elements of migrant groups, especially in the UK context, have been limited. This study contributes to that gap by demonstrating how Syrians in Liverpool actively engage in traditional practices, such as home decoration, household items, and social rituals, to maintain their cultural identity.

Ahmed (1999) viewed home-making as a dynamic negotiation of cultural practices within new contexts [61]. Hadjiyanni's (2007) research on Somalian refugees in the USA found that migrants creatively adapted their living environments to replicate familiar cultural elements within standardized housing units. Techniques included using bold patterns, traditional incense, and multiple layers of drapery to recreate native landscapes [62]. Fifić (2023) similarly found that Afghan refugees in the USA incorporated traditional furniture and textiles to mirror their homes in Afghanistan, using elaborately embroidered cushions and richly colored carpets to ensure comfort and familiarity [63]. This aligns with Li's (2016) study of Chinese migrants, who argued that interior design is closely linked to cultural history, with design forms shaped by tradition [64]. Wang (2021) similarly found that elderly immigrants conceptualize home as a journey of maintaining family [65]. This study contributes to these emerging discussions by detailing how Syrian families in the UK use traditional crafts, decorations, and scents to recreate their homeland in smaller spaces. Solutions such as building conservatories and reconfiguring rooms highlight the adaptive nature of their home-making practices. Incorporating culturally meaningful items into home decor serves as a tangible connection to their heritage, providing psychological comfort. The maintenance of special cabinets for traditionally crafted items is particularly symbolic, acting as focal points for family gatherings and preserving cultural traditions.

Researchers like Ager and Strang (2008) and Spencer (2011) have explored the role of social practices in fostering community and integration [66,67]. This study builds on this work by emphasizing the importance of hospitality and communal support among Syrians in the UK. The continuation of social rituals, such as Ajer (grieving practices) and the celebration of Eid al-Fitr and Ramadan, strengthens intra-community bonds while also promoting intercultural exchange with broader UK society. These findings align with the idea that social practices are central to the integration process, fostering mutual respect and understanding between different cultural groups.

The transmission of cultural heritage to younger generations has been another key theme in the literature, with Levitt (2009) [68] and Portes and Rumbaut (2001) [69] emphasizing the role of community-based education in preserving cultural practices. This study highlights the establishment of Arabic weekend schools by Syrians in the UK as a crucial method of cultural transmission, helping children stay connected to their heritage while also attracting interest from other Arabic-speaking nationalities.

Finally, when asked about the items they brought with them upon arriving in the UK, Syrians emphasized the immense sentimental value of these objects, which serve as reminders of home and identity. This echoes Howard's (2003) observation that the emotional significance of objects is often not realized until they are lost [70]. Items like photograph albums or family heirlooms, while not necessarily valuable in monetary terms, are cherished for their sentimental value. These findings suggest that the connection between home, ICH, and identity requires further exploration, presenting a promising direction for future studies.

4.5. Strengths and Limitations

The existing literature on the ICH of migrants primarily focuses on Europe and other parts of the world, with limited attention given to the UK context [71,72]. Although a few studies address the UK, they tend to concentrate on broader aspects of culture and place-making rather than specific elements of ICH, such as traditional crafts, objects, and decorations [73,74]. This study fills this gap by examining how Syrian families in the UK use ICH elements, including household items and scents, to recreate a sense of their homeland. These tangible connections to heritage provide psychological comfort and reinforce community bonds. This research contributes to the broader discourse on ICH and migration, highlighting the unique bottom-up approach, with interviews conducted entirely in Arabic, capturing the nuances of Syrian home-making practices in the UK. By exploring the motivations for representing their homes in ways that resemble traditional Syrian dwellings, this study showcases the resilience of the community. The ability to maintain Syrian traditions within physical surroundings not only aids in preserving cultural identity but also supports the community's recovery from war and displacement. One key strength of this study is the rich qualitative data obtained through direct access to the participants' homes, offering deep insights into their adaptive strategies. However, there are several limitations.

The relatively small sample size of 22 interviewees, while offering a foundational understanding of the ICH practices, may not fully represent the wider Syrian community in the UK. Furthermore, all interviewees were Muslims, which limits the generalizability to other religious Syrian groups. Compared to similar studies, the sample size is moderate. For example, Robertson-Rose (2022) focused on 25 Afghan refugees in the UK [75], and Baillie and Chippindale (2006) interviewed 35 participants across migrant communities [76]. Qualitative research does not aim for generalization but rather seeks to provide detailed, context-specific data to inform future research [77,78]. Another limitation is the geographical focus on Liverpool, which has a unique history as a multicultural port city. Existing support networks and local policies in Liverpool may differ from other UK cities, influencing Syrians' experiences with integration and home-making. Thus, while many of the identified themes—such as the role of ICH in home-making and the emotional importance of preserving traditions—are consistent with studies on other refugee groups, they may vary depending on the local context.

Lastly, this study highlights the need for further exploration of the relationship between ICH, cultural heritage policies, refugee policies, and home-making. A more extensive sample and additional research are necessary to explore how international and national policies could better align with refugee protection and cultural preservation, as discussed by Al Shallah (2024) [79].

4.6. Policy Implications and Future Research

This study highlights the importance of supporting refugees in their efforts to preserve and transmit cultural heritage, particularly through home-making practices. Policymakers and refugee support organizations should consider the cultural and emotional needs of refugee communities in addition to their physical needs. Providing resources that enable refugees to acquire culturally meaningful household items and create spaces for communal gatherings could significantly enhance their sense of belonging and integration. There is also a potential for an updated international cultural heritage law to safeguard the refugees' intangible cultural heritage (ICH), not only in terms of home practices and associations but also in relation to the knowledge and representations of the refugee experience within each community. A similar approach could be applied to protect the refugees' tangible cultural heritage. Such discussions are especially relevant now that the UK has recently ratified the 2003 UNESCO Convention on Safeguarding Intangible Cultural Heritage. This raises important questions about the inclusion of the resettled communities' ICH elements within national heritage inventories.

Future research should explore how subsequent generations of Syrians maintain and adapt ICH practices and whether these traditional practices evolve in response to new surroundings. Additionally, studies on the role of digital platforms in preserving ICH among displaced populations could offer insights into how cultural memory is sustained in the absence of physical objects.

5. Conclusions

This study underscores the resilience of the Syrian community in Liverpool by illustrating how their traditional cultural practices and household arrangements play a pivotal role in fostering a sense of home and belonging in displacement. The integration of tangible heritage, such as crafted objects and household items, alongside intangible cultural practices, like storytelling, religious rituals, and hospitality customs, highlights the interplay between material culture and ICH. These elements, when combined, serve as carriers of memory and identity, ensuring cultural continuity amidst new environments. By actively engaging in traditional crafts and maintaining cultural rituals, Syrians in Liverpool have not only recreated a sense of home but also demonstrated how ICH can guide adaptive and sustainable design practices in diaspora communities. This research responds to the key themes in the journal's special issue by showing how cultural heritage can influence the built environment, particularly within the context of migration and displacement.

This study makes an original contribution to the growing interdisciplinary discourse on ethnography, migration, home, place-making, and ICH by providing insight into the specific home-making practices of Syrians in Liverpool. It addresses a gap in the literature concerning the intersection of ICH and home-making among forced migrants. By grounding the research in tangible objects and intangible traditions, the study enhances our understanding of how displaced communities use cultural heritage as a coping mechanism and a means of sustaining their identity in the face of forced migration. Furthermore, it sheds light on the emotional and cultural significance of domestic spaces, which serve as vital loci of memory, continuity, and belonging for refugee communities.

While this study focuses on Syrians in Liverpool, further research could expand to compare the home-making practices of Syrians or other migrant communities in different regions of the UK and globally. Comparative studies across cities with varying levels of support systems, housing policies, and cultural diversity could provide deeper insights into how context influences the preservation of ICH and the home-making process. Moreover, future research could explore how subsequent generations of Syrians in the UK navigate the transmission of ICH in domestic spaces and whether traditional practices evolve or diminish over time in response to their new surroundings. Additionally, further studies could explore the gender dynamics of home-making among forced migrants, examining how men and women may engage differently with ICH in their homes. Another avenue for future research is the role of digital platforms and online communities in preserving ICH among displaced populations, particularly in cases where tangible objects may be lost or unavailable.

Overall, this research emphasizes the need for a deeper understanding of how cultural heritage can aid in the adaptation and integration of migrant communities, ultimately enhancing their sense of identity and belonging in new environments. By offering a nuanced look at the relationship between ICH and home-making, this study contributes valuable insights to both the academic discourse and policy discussions surrounding refugee integration and cultural preservation.

Funding: This research was funded by the University of Liverpool Higher Education Innovation Fund (HEIF), funding reference number: 174398, with additional support from the University of Liverpool's Research Development Initiative Fund (RDIF).

Institutional Review Board Statement: This research was funded by the University of Liverpool, and ethical approval was obtained from the University of Liverpool's Research Ethics Committee (reference: 11100).

Informed Consent Statement: Informed consent was obtained from all subjects involved in the study.

Data Availability Statement: The datasets presented in this article are not readily available due to interviewees privacy and adherence to ethical restrictions.

Acknowledgments: Photographs of the activities were skillfully captured by Monika Koeck, to whom I express my deepest gratitude. I also extend my heartfelt thanks to the Syrian British Cultural Centre in Liverpool for their unwavering support and to all the Syrians who generously participated in this project. A special acknowledgement goes to the Syria Trust for Development for providing invaluable materials that significantly enriched the project's outcomes. Lastly, I am profoundly grateful to the University of Liverpool for their generous funding and unwavering support, which made this project possible.

Conflicts of Interest: The author declares no conflicts of interest.

References

1. Lozanovska, M. *Migrant Housing: Architecture, Dwelling, Migration*, 1st ed.; Routledge: London, UK, 2019. [CrossRef]
2. Statista. Number of Syrian Nationals Residing in the United Kingdom from 2008 to 2021. Available online: https://www.statista.com/statistics/1253266/syrian-population-in-united-kingdom/#:~=There%20were%20approximately%2028%20thousand,2019%20with%2047%20thousand%20nationals (accessed on 5 July 2024).
3. Home Office. *Syrian Vulnerable Persons Resettlement Scheme (VPRS) Guidance for Local Authorities and Partners*; Crown Copyright: London, UK, 2017.
4. UNHCR. Resettlement in the United Kingdom. Available online: https://help.unhcr.org/uk/resettlement/ (accessed on 1 June 2024).
5. Beeckmans, L.; Gola, A.; Singh, A.; Heynen, H. (Eds.) *Making Homes in Displacement: Critical Reflections on a Spatial Practice*; Springer: Berlin, Germany, 2022.
6. UNESCO. What is Intangible Cultural Heritage? Available online: https://ich.unesco.org/en/what-is-intangible-heritage-00003 (accessed on 5 July 2024).
7. Syria Trust for Development. Intangible Cultural Heritage Elements. Available online: https://www.syriatrust.sy/en/affiliate/14 (accessed on 5 July 2024).
8. Sokolova, A.N. The Caucasian-Scottish Relations Through the Prism of the Fiddle and Dance Music. In Proceedings of the North Atlantic Fiddle Convention, Aberdeen, Scotland, 26–30 July 2006; The Elphinstone Institute, University of Aberdeen: Aberdeen, Scotland, 2006.
9. ICH of Syria. Available online: https://ich.sy/ar/side/menu/content/3 (accessed on 25 June 2024).
10. Syrian Heritage Archive Project. Available online: https://syrian-heritage.org/ajami-or-damascene-painting-traces-of-a-traditional-handcraft/ (accessed on 5 September 2024).

11. Easthope, H. A Place Called Home. *House Theory Soc.* **2004**, *21*, 128–138. [CrossRef]
12. Smyth, M.; Croft, S. *Culture and Sustainability: Spaces, Materials and Aesthetic Considerations*; Springer: Berlin, Germany, 2006.
13. Perkins, H.C.; Thorns, D.C.; Newton, B.M. The study of home from a social scientific perspective. *Hous. Stud.* **2002**, *17*, 805–823.
14. Aragonés, J.I.; Amérigo, M.; Pérez-López, R. Perception of personal identity at home. *Psicothema* **2010**, *22*, 872–879. [PubMed]
15. Dossa, P.; Golubovic, J. *Re-Imagining Home: An Ethnographic Perspective*; Routledge: London, UK, 2019.
16. Brun, C.; Fabos, A. Making Homes in Limbo? A Conceptual Framework. *Refuge* **2015**, *31*, 5–17. [CrossRef]
17. Kreuzer, M.; Mühlbacher, H.; von Wallpach, S. Home in the re-making: Immigrants' transcultural experiencing of home. *J. Bus. Res.* **2018**, *91*, 334–341. [CrossRef]
18. Levin, I. *Migration, Settlement, and the Concepts of House and Home*; Routledge: London, UK, 2015.
19. Motasim, H.; Heynen, H. At home with displacement? Material culture as a site of resistance in Sudan. *Home Cult.* **2011**, *8*, 43–69. [CrossRef]
20. Boccagni, P.; Brighenti, A. Immigrants and home in the making: Thresholds of domesticity, commonality and publicness. *J. Hous. Built Environ.* **2015**, *30*, 561–579. [CrossRef]
21. UNHCR. #WhatHomeMeans Campaign. Available online: https://www.unrefugees.org.uk/whathomemeans/ (accessed on 1 April 2024).
22. Şenoğuz, P.; Carpi, E. Refugee Hospitality in Lebanon and Turkey. On Making 'The Other'. *Int. Migr.* **2018**, *57*, 92.
23. Ilcan, S.; Rygiel, K.; Baban, F. The Ambiguous Architecture of Precarity: Temporary Protection, Everyday Living, and Migrant Journeys of Syrian Refugees. *Int. J. Migr. Bord. Stud.* **2018**, *4*, 10012236. [CrossRef]
24. Yalçın, M.G.; Düzen, N.E. A Paradox of Inclusion in Cities: Homemaking of Refugees and Natives in Urban Space. *Int. Migr. Integr.* **2024**, 1–24. [CrossRef]
25. Albadra, D.; Coley, D.; Hart, J. Toward healthy housing for the displaced. *J. Archit.* **2018**, *23*, 115–136. [CrossRef]
26. Kirk, D.S.; Nabil, S.; Talhouk, R.; Trueman, J.; Bowen, S.; Wright, P. Decorating public and private spaces: Identity and pride in a refugee camp. In Proceedings of the Conference on Human Factors in Computing Systems, Montreal, QC, Canada, 21–26 April 2018. [CrossRef]
27. Paszkiewicz, N.; Fosas, D. Reclaiming refugee agency and its implications for shelter design in refugee camps. In Proceedings of the International Conference on Comfort at the Extremes: Energy, Economy and Climate, Dubai, United Arab Emirates, 10–11 April 2019; pp. 584–594.
28. Dalal, A. Uncovering Culture and Identity in Refugee Camps. *Humanities* **2017**, *6*, 61. [CrossRef]
29. Zibar, L.; Abujidi, N.; de Meulder, B. Who/What Is Doing What? Dwelling and Homing Practices in Syrian Refugee Camps—The Kurdistan Region of Iraq. In *Making Homes in Displacement: Critical Reflections on a Spatial Practice*; Beeckmans, L., Gola, A., Singh, A., Heynen, H., Eds.; Springer: Berlin, Germany, 2022.
30. Doná, G. Making homes in limbo: Embodied virtual "homes" in prolonged conditions of displacement. *Refuge* **2015**, *31*, 67. [CrossRef]
31. Korac, M. *Remaking Home: Reconstructing Life, Place and Identity in Rome and Amsterdam*; Berghahn Books: New York, NY, USA; Oxford, UK, 2009. [CrossRef]
32. Arvanitis, E.; Yelland, N. 'Home means everything to me. . .': A study of young Syrian refugees' narratives constructing home in Greece. *J. Refug. Stud.* **2021**, *34*, 535–554. [CrossRef]
33. Shamma, Y.; Ilcan, S.; Squire, V.; Underhill, H. (Eds.) *Migration, Culture and Identity: Making Home Away*; Springer: Berlin, Germany, 2022.
34. Hashem, R.; Dudman, P.V.; Shaw, T. Archiving Displacement and Identities: Recording Struggles of the Displaced, Remaking Home in Britain. In *Migration, Culture and Identity, Politics of Citizenship and Migration*; Shamma, Y., Ilcan, S., Squire, V., Underhill, H., Eds.; Springer: Berlin, Germany, 2022.
35. Rifeser, J.; Puntil, D.; Borelli, E. (Un)doing home: Exploring home-making and identity—An example project in a London higher education classroom. *Lond. Rev. Educ.* **2023**, *21*, 33. [CrossRef]
36. Travlou, P. From Cooking to Commoning: The Making of Intangible Cultural Heritage in OneLoveKitchen, Athens. In *Cultural Heritage in the Realm of the Commons: Conversations on the Case of Greece*; Lekakis, S., Ed.; Ubiquity Press: London, UK, 2020; pp. 159–182. [CrossRef]
37. New Lines Institute. Rohingya Cultural Preservation: An Internationally Coordinated Response Is Urgent. Available online: https://newlinesinstitute.org/state-resilience-fragility/complex-emergencies-and-humanitarian-crises/rohingya-cultural-preservation-an-internationally-coordinated-response-is-urgent/ (accessed on 11 November 2024).
38. Homeland Document. Oral History Recordings Projects in Syria. Available online: https://wathiqat-wattan.org/category/oral-history (accessed on 1 June 2024).
39. Muhesen, N. Syrian Intangible Cultural Heritage: Characteristics and Challenges of Preservation. In *Tangible and Intangible Heritage in the Age of Globalisation*; Makhlouf, L., Ed.; Open Book Publishers: Cambridge, UK, 2024; pp. 89–100. [CrossRef]
40. Burrell, K.; Lawrence, M.; Wilkins, A. *At Home in Liverpool During COVID-19*; Queen Mary University of London: London, UK; University of Liverpool: Liverpool, UK, 2021.
41. Belchem, J. *Liverpool 800: Culture, Character and History*; Liverpool University Press: Liverpool, UK, 2006; pp. 9–58.
42. Rishbeth, C.; Ganji, F.; Vodicka, G. Ethnographic understandings of ethnically diverse neighbourhoods to inform urban design practice. *Local Environ.* **2017**, *23*, 36–53. [CrossRef]

43. Gentner, D.; Holyoak, K.J.; Kokinov, B.N. (Eds.) *The Analogical Mind: Perspectives from Cognitive Science*; MIT Press: Cambridge, MA, USA, 2001.

44. Holyoak, K.J.; Thagard, P. *Mental Leaps: Analogy in Creative Thought*; MIT Press: Cambridge, MA, USA, 1995.

45. Kvale, S. *Interviews: An Introduction to Qualitative Research Interviewing*; Sage: London, UK, 1996.

46. Brinkmann, S.; Kvale, S. *Doing Interviews*, 2nd ed.; SAGE Publications Ltd.: London, UK, 2018.

47. White, M.D.; Marsh, E.E. Content analysis: A flexible methodology. *Libr. Trends* **2006**, *55*, 22–45. [CrossRef]

48. Saldaña, J. *The Coding Manual for Qualitative Researchers*, 3rd ed.; SAGE: Los Angeles, CA, USA, 2016.

49. Sparkes, A.C.; Smith, B. *Narrative Analysis as an Embodied Engagement with the Lives of Others*; Sage Publications: Newbury Park, CA, USA, 2012.

50. Angrosino, M. *Doing Ethnographic and Observational Research*; Sage Publications: London, UK, 2007.

51. Creswell, J.W. *Qualitative Inquiry and Research Design: Choosing Among Five Approaches*; Sage Publications: Los Angeles, CA, USA, 2013.

52. Spradley, J.P. *Participant Observation*; Holt, Rinehart, and Winston: New York, NY, USA, 1980.

53. Lofland, J.; Lofland, L.H. *Analyzing Social Settings: A Guide to Qualitative Observation and Analysis*; Wadsworth Publishing: Belmont, CA, USA, 1995.

54. Emerson, R.M.; Fretz, R.I.; Shaw, L.L. *Writing Ethnographic Fieldnotes*; University of Chicago Press: Chicago, IL, USA, 2011.

55. Boccagni, P. *Migration and the Search for Home: Mapping Domestic Space in Migrants' Everyday Lives*; Palgrave Macmillan: London, UK, 2017.

56. Ilcan, S.; Squire, V. Migratory Journeys, State Refugee Policies, and Negotiated Belonging. In *Migration, Culture and Identity, Politics of Citizenship and Migration*; Shamma, Y., Ilcan, S., Squire, V., Underhill, H., Eds.; Springer: Berlin, Germany, 2022; pp. 123–146.

57. Baser, B. Kurdish Diaspora Political Activism in Europe with a Particular Focus on Great Britain. Ph.D. Thesis, University of Exeter, Exeter, UK, 2011.

58. White, A. *Polish Families and Migration Since EU Accession*; Policy Press: Bristol, UK, 2015; pp. 1–224.

59. Hapsariniaty, A.W.; Darmaningtyas, P.; Subagio, I.; Pratiwi, W.D.; Rahadi, R.A. Culture and Domestic Spaces: The Influence of Culture on the Interior Residential Setting. *J. Soc. Sci. Res.* **2019**, *5*, 1816–1827. [CrossRef]

60. Momade, Y. The Influence of Culture on Interior Design—Morocco: From Case Studies and Expert Interviews. Master's Thesis, IADE—Faculdade de Design, Tecnologia e Comunicação da Universidade Europeia, Lisbon, Portugal, 2022.

61. Ahmed, S. Home and Away: Narratives of Migration and Estrangement. *Int. J. Cult. Stud.* **1999**, *2*, 329–347. [CrossRef]

62. Hadjiyanni, T. Bounded Choices: Somali Women Constructing Difference in Minnesota Housing. *J. Inter. Des.* **2007**, *32*, 13–27. [CrossRef]

63. Fifić, A. Forced Immigrants and Placemaking: How the Interior Design of Refugee Homes Impacts Their Smooth Integration in Host Community. Master's Thesis, University of Oklahoma, Norman, Oklahoma, 2023.

64. Li, W. The Inheritance and Development of Traditional Culture in Interior Design and Three Dimensional Structure. In Proceedings of the 2nd International Conference on Economics, Social Science, Arts, Education and Management Engineering (ESSAEME 2016), Huhhot, China, 30–31 July 2016.

65. Wang, Q.; Zhan, H.J. The Making of a Home in a Foreign Land: Understanding the Process of Home-Making Among Immigrant Chinese Elders in the U.S. *J. Ethn. Migr. Stud.* **2021**, *47*, 1539–1557. [CrossRef]

66. Ager, A.; Strang, A. Understanding Integration: A Conceptual Framework. *J. Refug. Stud.* **2008**, *21*, 166–191. [CrossRef]

67. Spencer, S. *The Migration Debate*; Policy Press: Bristol, UK, 2011; pp. 1–272.

68. Levitt, P. Roots and Routes: Understanding the Lives of the Second Generation Transnationally. *J. Ethn. Migr. Stud.* **2009**, *35*, 1225–1242. [CrossRef]

69. Portes, A.; Rumbaut, R.G. *Legacies: The Story of the Immigrant Second Generation*; University of California Press: Berkeley, CA, USA, 2001; pp. 1–396.

70. Howard, P. *Heritage: Management, Interpretation, Identity*; Bloomsbury Publishing Plc: London, UK, 2003.

71. Giglitto, D.; Ciolfi, L.; Bosswick, W. Building a bridge: Opportunities and challenges for intangible cultural heritage at the intersection of institutions, civic society, and migrant communities. *Int. J. Herit. Stud.* **2022**, *28*, 74–91. [CrossRef]

72. Naguib, S. Museums, Diasporas, and the Sustainability of Intangible Cultural Heritage. *Sustainability* **2013**, *5*, 2178–2190. [CrossRef]

73. Gedalof, I. Unhomely Homes: Women, Family and Belonging in UK Discourses of Migration and Asylum. *J. Ethn. Migr. Stud.* **2007**, *33*, 77–94. [CrossRef]

74. Helen, T. Refugees, the State and the Concept of Home. *Refug. Surv. Q.* **2013**, *32*, 130–152. [CrossRef]

75. Robertson-Rose, E. What is Home? Experiences of Home After Forced Migration from Afghanistan to the UK as a Refugee. Master's Thesis, Aalborg University, Aalborg, Denmark, 2022.

76. Baillie, B.; Chippindale, C. Tangible–Intangible Cultural Heritage: A Sustainable Dichotomy? *Conserv. Manag. Archaeol. Sites* **2006**, *8*, 174–176. [CrossRef]

77. Smith, J.A. *Qualitative Psychology: A Practical Guide to Research Methods*; SAGE Publications: Thousand Oaks, CA, USA, 2018.

78. Denzin, N.K.; Lincoln, Y.S. *The SAGE Handbook of Qualitative Research*; SAGE Publications: Thousand Oaks, CA, USA, 2011.
79. Al Shallah, S. Refugee Protection through Safeguarding Intangible Cultural Heritage of the Home Country and Refugee Journey. *Int. J. Cult. Prop.* **2024**, *30*, 280–297. [CrossRef]

Disclaimer/Publisher's Note: The statements, opinions and data contained in all publications are solely those of the individual author(s) and contributor(s) and not of MDPI and/or the editor(s). MDPI and/or the editor(s) disclaim responsibility for any injury to people or property resulting from any ideas, methods, instructions or products referred to in the content.

MDPI AG
Grosspeteranlage 5
4052 Basel
Switzerland
Tel.: +41 61 683 77 34

Architecture Editorial Office
E-mail: architecture@mdpi.com
www.mdpi.com/journal/architecture

Disclaimer/Publisher's Note: The title and front matter of this reprint are at the discretion of the Guest Editor. The publisher is not responsible for their content or any associated concerns. The statements, opinions and data contained in all individual articles are solely those of the individual Editor and contributors and not of MDPI. MDPI disclaims responsibility for any injury to people or property resulting from any ideas, methods, instructions or products referred to in the content.

www.ingramcontent.com/pod-product-compliance
Lightning Source LLC
LaVergne TN
LVHW071936160726
843515LV00011B/2628